The Cambridge Manuals of Science and
Literature

THE MODERN WARSHIP

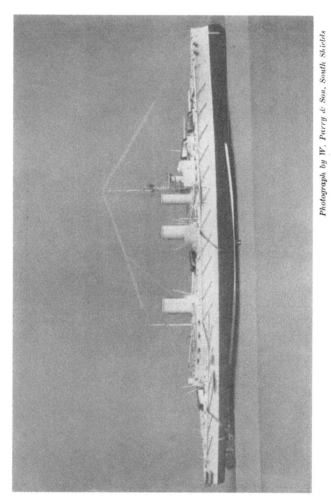

Photograph by W. Parry & Son, South Shields

Finished Model of H.M.S. *Queen Mary*

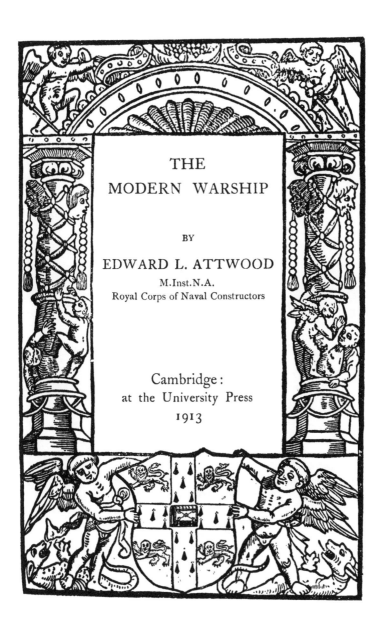

THE
MODERN WARSHIP

BY

EDWARD L. ATTWOOD

M.Inst.N.A.
Royal Corps of Naval Constructors

Cambridge:
at the University Press
1913

CAMBRIDGE UNIVERSITY PRESS
Cambridge, New York, Melbourne, Madrid, Cape Town,
Singapore, São Paulo, Delhi, Tokyo, Mexico City

Cambridge University Press
The Edinburgh Building, Cambridge CB2 8RU, UK

Published in the United States of America by
Cambridge University Press, New York

www.cambridge.org
Information on this title: www.cambridge.org/9781107401778

First published 1913
First paperback edition 2011

A catalogue record for this publication is available from the British Library

ISBN 978-1-107-40177-8 Paperback

*With the exception of the coat of arms
at the foot, the design on the title page is a
reproduction of one used by the earliest known
Cambridge printer, John Siberch, 1521*

'For one thing this century will in after years be considered to have done in a superb manner and one thing I think only.

'It will always be said of us with unabated reverence *They built ships of the line.* Take it all in all a ship of the line is the most honourable thing that man as a gregarious animal has ever produced. By himself, unaided, he can do better things than ships of the line, he can make poems and pictures, and other combinations of what is best in him.

'But as a being living in flocks and hammering out with alternate strokes and mutual agreement what is necessary for him in those places to get or produce, the ship of the line is his first work. Into that he has put as much of his human patience, common sense, forethought, experimental philosophy, self control, habits of order and obedience, thoroughly wrought hand-work, defiance of brute elements, careless courage, careful patriotism and calm expectation of the judgment of God as can well be put into a space of 300 feet long and 80 feet broad and I am thankful to have lived in an age when I can see this thing done.'

JOHN RUSKIN,

Harbours of England.

PREFACE

IT is hoped that the brief account of the modern warship contained in the following pages, written from the naval architect's point of view, may be found of interest and service to the general reader who desires to obtain some acquaintance with the subject.

Limitations of space have prevented reference being made to many points of interest in connection with warship design and construction, and anything of a controversial or confidential nature has been avoided. Also no reference has been made to contemporary warships of other countries; these are fully dealt with, so far as the facts are available, in the *Naval Annual* and similar publications. Figs. 7, 12, 13, 14, and 16 have been taken from *Warships*, by permission of Messrs Longmans, Green & Co., the publishers.

The frontispiece, a photograph of the finished model of H.M.S. *Queen Mary*, has been included by permission of Palmer's Shipbuilding and Iron Co., Ltd., Jarrow-on-Tyne. This model has been graciously accepted by Her Majesty the Queen.

The naval architect who is responsible for the completed ship must necessarily obtain the co-operation of specialists in many departments. Among these may be mentioned the marine, electrical and hydraulic engineers, the artillerist, the metallurgist, the torpedo and compass specialists and the naval officers who have to navigate, work, and fight the ship. It is only by loyal co-operation that a satisfactory result can be achieved. A warship costing, say, two millions sterling is required to be built, launched, armoured, fitted out with machinery, armament, etc., and tested by speed, gun and torpedo trials in a period of time aggregating only about 600 working days. At the beginning of this time the vessel is a design on paper, at the end of this time she is a completed unit of the British Navy ready and fit 'to go anywhere and do anything.' Such a performance requires not only great forethought, skill and ability, but also admirable organisation and co-ordination between workers in many departments, and, it is submitted, deserves to be placed in the front rank of the engineering achievements of our time.

E. L. A.

London, *January* 1913.

CONTENTS

CHAPTER I

DESIGN

THERE are two distinct types of the modern war-ship of large size, viz.:—the battleship and the battle-cruiser. The former has thicker armour, more guns and less speed than the latter. The calibre of the guns now employed in these types of vessel is the same and the great speed of the battle-cruiser is obtained at the sacrifice of armour protection and number of guns. This distinction may be illustrated by the comparison on page 2 between the *Hercules* and *Indefatigable* and between the *Orion* and the *Lion*, in regard to which particulars have been officially published.

The capital on which a designer has to work is the displacement or total weight available and for a given displacement any increase in one feature of a design must be compensated for by a decrease in another feature or other features. If the weights allowed in a design are exceeded in the aggregate in the completed ship, there must be the inevitable penalty of increase in draught over that originally

	Hercules Battleship	Indefatigable Battle-cruiser	Orion Battleship	Lion Battle-cruiser
Date of Launch	1910	1909	1910	1910
Length	510'	555'	545'	660'
Breadth	85'	80'	88½'	88½'
Draught	27'	26½'	27½'	28'
Coals at this draught	900 tons	1000 tons	900 tons	—
Displacement	20,000 tons	18,750 tons	22,500 tons	26,350 tons
Horse-Power	25,000	43,000	27,000	70,000
Speed Designed. Knots	21	25	21	28
Weight of Hull and Armour	12,440 tons	10,735 tons	—	—
Main Armament	10 12-in.	8 12-in.	10 13½-in.	8 13½-in.
Minor Armament	16 4-in.	16 4-in.	16 4-in.	16 4-in.
Torpedo Tubes	3 21-in.	3 18-in.	3 21-in.	2 21-in.
Max. thickness of Belt Armour	11 in.	7 in.	12 in.	9 in.
Total cost excluding stores	£1,660,950	£1,536,769	£1,918,773	£2,068,337

(A knot is 1¼ miles an hour, so that 28 knots is 32 miles an hour.)

allowed for. The details of the weights of ships of the Royal Navy are not published by the British Admiralty but the following percentages have been given by Sir Philip Watts for a 1905 battleship, viz.:

Equipment	...	...	4%	...	(655 tons)
Armament	...	...	19%	...	(3110 ,,)
Propelling machinery		...	10·5%	...	(1720 ,,)
Coal	...	...	...	5·5%	...	(900 ,,)
Armour	...	...	...	26%	...	(4250 ,,)
Hull	...	...	...	35%	...	(5730 ,,)

which lead to the figures in brackets for a ship of displacement 16,365 tons.

It will have been noticed in the above figures for battleships that a weight of 900 tons is included for coal. This is described as the 'legend' weight and is employed purely for convenience, that is to say, it is the weight of coal assumed for design purposes, at the draught of water stated, at which draught the official speed trials are carried out. The total coal capacity is always more than this, e.g. in the *Dreadnought* this is 2700 tons and besides this the spaces between the inner and outer skins are arranged to carry oil fuel. The addition of weights beyond those allowed in the 'legend' figures of course increases the draught and reduces the speed.

The design of a war-vessel of novel type is necessarily a matter of trial and error. The various portions of the design are interdependent. Thus,

suppose a certain speed is desired; the horse-power
for this speed is dependent on the size and form of
the ship which at the outset are unknown. It is
necessary to proceed by a series of approximations,
and to produce a tentative design, which may or may
not satisfy all the required conditions, but which may
be used as a starting-point for alterations. In some
cases the size of the ship may be limited by the provision available for dry docking or a limit may be set
on the total cost. The design of the form of the
ship most suitable for the desired speed is assisted
materially by experiments on small scale models in
an experimental tank. Speed and length of ship have
an important connection. It is not possible in full-
sized ships to obtain high speeds on moderate lengths
because of the excessive horse-power necessary.
This is the reason for the greater length of modern
cruisers of 25 to 28 knots speed as compared with
battleships of 21 knots speed. This point will be
referred to later, but it may be stated here that the
ratio of the speed to the square root of the length
of ship must not exceed about 1·1 or an excessive
expenditure of power is necessary. When the length
and form of ship are tentatively decided the estimate
of power can be made based on previous experience
and confirmed by model experiments. This power will
lead to a certain weight and space for the machinery.
The weight of hull can then be approximated to, as

also that of the armament. The difference between
the aggregate of these weights and the total displacement will give the weight available for stores, equipment, coal, and protection. If this weight is not
sufficient some revisions in the design will be necessary.
By means of this process of trial and error a balance
is ultimately obtained between the grand total of the
weights and the displacement corresponding to the
size and form of the ship. When a new design is
simply a variation of a previous design the work is
rapidly performed, but when novel conditions depart-
ing considerably from precedent have to be met, the
ultimate result is only obtained by considering many
alternatives.

There are two essential conditions which must be
fulfilled in any design, viz.:—sufficient strength and
sufficient stability. The strength of a ship has to be
both 'structural' and 'local.' Structural strength is
the strength of the ship regarded as a complete
structure. Local strength is that of special portions.

The most important structural strains are those
due to bending in a fore and aft direction, the ship
being regarded as a huge beam or girder. When a
Civil Engineer designs a bridge he can obtain a fairly
close approximation as to the strains to which his
structure is likely to be subject. He has fixed
supports, and calculations can be made as to weights,
loads, wind pressures, etc. In the case of a ship,

however, such exact calculations cannot be made
because of the indeterminate nature of the conditions
at sea to which a ship may be subject. All that can
be done is to make certain extreme assumptions as
to the position of the ship on a wave and the strength
calculations are based on these assumptions. This
method has been adopted for many years with
satisfactory results. Two extreme conditions are
assumed, viz.:

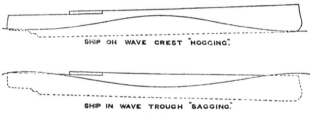

Fig. 1.

(*a*) the ship is supposed to be momentarily
poised on the crest of a wave having the same length
as the ship (as fig. 1), and

(*b*) the ship is supposed to be momentarily
across the trough of a similar wave (as fig. 1).

The wave in each case is usually assumed to have
a height equal to one-twentieth its length. The first
condition is termed '*hogging*,' the ends of the ship
tending to drop relatively to the middle, the upper

portions of the structure tending to tear apart, and
the lower portions to buckle. The second condition
is termed '*sagging*,' the upper portions tending to
buckle and the lower portions to tear apart. The
distribution of the weights of the ship has first to be
determined and then the distribution of the upward
support of the buoyancy on the assumptions (*a*) and
(*b*) above : each of these leads to a definite bending
moment on the midship structure. By means of the
ordinary beam formula a measure of the stresses on
the upper and lower portions of the structure can be
calculated, and these stresses are compared with
those of completed ships, obtained by a similar
process of calculation, which have proved satisfactory
on service.

The maximum bending moment is found to vary
as the product of the weight and the length and may
be expressed approximately by the formula;

Bending moment = coefficient × W × L where W
is the total weight or displacement and L is the
length.

Thus assuming that the coefficient is the same in
both cases the bending moment in a ship like the
Lion (battle-cruiser of 660 feet length and 26,350 tons
displacement) will be about 70 % greater than that of
the *Indefatigable* (battle-cruiser of 555 feet length
and 18,750 tons displacement) and the structure of
the former ship must be increased in comparison

with that of the latter ship so as to obtain a similar
value for the maximum stress on the material com-
posing the section.   This illustrates very forcibly the
great increase of bending strains that has to be
provided for as ships increase in size.

The transverse strength of a ship is also of importance.   When a ship is rolling heavily there is a
tendency to distortion in a transverse direction.   It
is a necessary condition of warship design to divide
the ship into a large number of separate watertight
compartments, and the numerous transverse divisions
or bulkheads tie the ship together and prevent any
transverse racking.   It is now the practice to make
the bottom of large warships, amidships, flat over as
large an area as possible and when docked such ships
sit down on three rows of blocks.

Local strains have to be provided for by special
arrangements.   Divisional bulkheads must be strong
enough to withstand the strains due to a compartment
on one side being flooded and the stiffening of these
bulkheads when of large area is of a specially strong
character.   The safety or control of a ship might
very conceivably depend on a main bulkhead remaining intact under such circumstances.   Strong girders
and pillars are arranged beneath the decks under the
blast of heavy guns.   Heavy weights, such as capstans
and auxiliary machinery, are supported by pillars and
strong girders.   A tremendous localised weight has

to be supported underneath a twin gun mounting. There is not only the weight of the fixed armoured barbette protecting the mounting, but the weight of the revolving turntable containing the gun mounting, guns and armoured shield, and the strains due to firing the guns. Beneath such places the framing of the ship is specially strengthened and strong bulkheads and pillars are arranged to distribute the strains throughout the structure. Special attention is necessary at the stern in order to provide against the strains due to the rudders and the propellers.

Proper stability is another essential in the design of a war vessel and provision has not only to be made for the case of the vessel intact and uninjured but also for a vessel which has sustained a reasonable amount of damage. Stability conditions are special in war vessels because of the heavy weights of armour, guns, and gun mountings carried at great heights above the water which lead to a high position of the centre of gravity of the ship. Stability is looked at from two points of view, viz.:—initial stability or stability near the upright condition and stability at large angles of inclination. A steady gun platform is also most desirable. The subject of stability and rolling will be dealt with in a later chapter.

Other features of a design, although secondary in importance to strength and stability, which have to be considered are :—

(1) *Armour protection to hull for preservation of buoyancy and stability.* This protection is specially important in the region of the waterline because a serious loss of stability is experienced if the area of the load water plane is diminished by the inflow of water due to damage. On this account the side

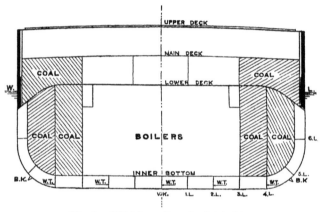

Fig. 2.   Midship section of a battleship.

compartments of a ship in this neighbourhood are well subdivided by watertight divisions so as to limit the amount of water entering supposing the armour were pierced. The armour protection is divided into (*a*) vertical armour and (*b*) thick decks, which are shewn in the section given in fig. 2. The lower deck is always sloped down to meet the lower edge of the

armour, the idea being that if a shot pierced the main
armour it would still have to penetrate the armour
deck before entering the vitals of the ship, or if a
shell burst inside the armour the deck would keep

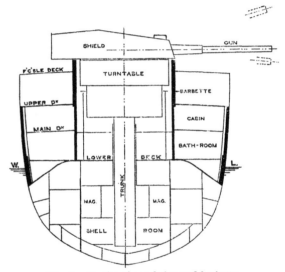

Fig. 3.   Section through forward barbette.

out the splinters. The upper armoured deck is fitted
at the upper edge of the armour.

(2) *Armour protection to armament.* This
usually consists of circular armoured barbettes which
protect the revolving turntable. The upper portion

of the gun mounting and the rear of the guns are
protected by an armoured shield which revolves with
the turntable and the guns (see fig. 3).

(3) *Armament*, consisting of guns and torpedoes.
All else is really subsidiary to this, as a warship is
essentially a gun platform and it is her function to
hit hard and hit often. The designer has so to arrange
the disposition of the armament that it shall be
well above water to give command and also to provide
for the most effective use of each portion of the
armament by giving the largest arcs of training
possible. A feature of recent designs has been that
every portion of the main armament can be trained
on either broadside.

(4) *Habitability*. A warship is the living place
of large numbers of men for long periods and it is
necessary to make such arrangements as shall conduce
to comfort and good health. The ventilation especially
has to be very carefully considered not only for fine
weather but also for foul weather when the ship must
be 'battened down.' The stowage of provisions,
messing, sleeping, washing and sanitary arrangements
all have to be considered in their turn and it is not
always an easy matter to provide satisfactory accom-
modation. It may be mentioned in this connection
that the standard of comfort on board ship has
steadily risen during the last twenty years or so as it
has on shore, and this has meant additions to weight

and space to what was formerly considered to be sufficient. As an example, stokers only used to be provided with wash-places, but the seamen now have wash-places set apart for their use. Formerly the sailors had to perform their ablutions in a tub somewhere on the deck, a practice which neither conduced to cleanliness nor decency.

(5) *Handiness.* A warship has to manœuvre in company with other ships and it is necessary that she shall respond quickly to the helm and turn in a small circle. Some remarks on the subject of steering and turning will be given in a later chapter.

(6) *Stowage of coal, oil fuel, and ammunition.* The length of time a warship will remain efficient as a fighting machine depends on how much coal, oil fuel, and ammunition she can carry. Coal is not only carried abreast the boilers but also above the protection deck (see fig. 2) and in the *Dreadnought* there is a capacity for 2700 tons of coal in the bunkers. The spaces between the inner and outer bottoms are also arranged to carry oil fuel and the radius of action of a ship, which goes away from her base with her full capacity of oil fuel, is enormously increased thereby. The ammunition consists of the propellant 'cordite' and the projectiles, and the number of rounds that can be carried is a matter of importance.

(7) *Machinery and boilers.* These depend on the horse power required, which in turn depends on

the speed.   Boilers are now always of the water-tube
type, i.e. the water is inside the tubes and the fire is
outside.   With these boilers the weight of water
necessary is much less than formerly.   Boilers are
either of the 'Yarrow' or of the 'Babcock & Willcox'
type.   As regards machinery, all large vessels of the
Royal Navy commencing with the *Dreadnought* have
turbine machinery.

Influence of an added weight on a design.   In
ship design it has to be borne in mind that increasing
the weight to be carried involves a very much greater
increase in the total displacement.   Suppose for instance in a completed design it is desired to increase
the armour protection, involving an added weight of
500 tons, say, and the speed is to be the same.   More
power will be required for the heavier ship, the
machinery will then take up more room and require
a larger complement and a longer ship.   The larger
ship will cost more in weight of hull.   The various
items of the design act and react upon one another
and we should find that 500 tons additional armour
protection would involve an increase of displacement
of the ship of over 1000 tons.   Years ago 37-ft. steam
pinnaces were supplied to ships of the Royal Navy,
56-ft. steam pinnaces being now carried in many ships.
The actual additional weights of boats is but 18 tons,
but the heavier boats require strong masts and derricks and powerful hydraulic or electric hoists for

lifting, together with special stowage. All this involves an additional weight of about 70 tons and, for a constant speed, an addition of about 150 tons to the displacement, as compared with what would have been necessary with lighter boats.

Mr Owens, of Vickers, Ltd., has given figures in a recent article for the cost of installing torpedo tubes in a given ship. Taking a design with no torpedo tubes, with all the space economically utilised, the addition of eight torpedo tubes with their necessary torpedoes and gear would only involve of themselves about 90 tons weight. To accommodate these tubes however would mean an addition of 50 feet on to the length of ship, and to maintain the same speed the aggregate addition to the design amounts to the large figure of 1800 tons. In this case there is not only added weight but added space to be provided.

If economies of weight can be effected in a given ship during building it means that other useful weights (say coal) can be carried at the designed draught or at the lighter draught a greater speed can be obtained. It is highly desirable therefore, that, consistent with proper provision of strength, economies in weight should be effected wherever possible. Small savings do not appear to be in themselves of much value but in the aggregate they will amount to a very considerable saving of weight. It is a case of taking care of the pennies and even of

the farthings.   One instance of this may be referred
to here.   Structural material is liable to rejection if
supplied heavier than demanded, the only latitude
allowed being 5 % below.   If there were no such
condition insisted upon and the weight as supplied
paid for, it would be to the interest of the manufacturer to roll heavy, and 5 % on the structural
material weighing 5000 tons amounts to 250 tons, a
weight worth having to increase the fighting or
endurance qualities of the ship.

Although war-vessels have greatly increased in size
during recent years the increase has not been so great
as in the case of the large Atlantic Liners.   The
comparison has been somewhat obscured because of
the different ways of expressing tonnage.   Warships
are described as having so many tons displacement
or total weight at legend draught but merchant
vessels are described as so many tons (gross register)
which is an empirical measure of internal capacity in
which 100 cubic feet is called a ton.   The *Olympic*
(and the ill-fated *Titanic*) is 45,000 tons gross register
tonnage while of 60,000 tons displacement. The largest
vessel now in commission in the British Navy is the
Lion, of 26,350 tons displacement, so that so far as
weight is concerned the *Olympic* is $2\frac{1}{4}$ times as large
as the *Lion*.   The comparative dimensions of these
vessels are as follows:

	Length	Breadth	Draught	Depth to top strength deck	Displacement	Horse-Power	Speed
Lion	660'	88½'	28'	54'	26,350 tons	70,000	28 knots
Olympic	850'	92½'	34½'	73'	60,000 tons	46,000	21 knots

The comparative sections of the *Hercules* (battle-ship), *Lion* (battle-cruiser) and the *Olympic* are shewn in fig. 4 which gives an idea of the relative proportions. The *Olympic* is 67 % longer than the *Hercules* and 29 % longer than the *Lion*.

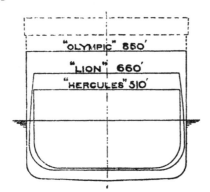

Fig. 4.   Comparative sections.

CHAPTER II

HULL CONSTRUCTION

THE major portion of the hull structure of a warship is formed of *mild steel*, a material of good tensile strength and great ductility. This mild steel has a tensile strength of 26 to 30 tons per square inch. In some ships, for important portions of the structure contributing to the longitudinal strength, as outer and inner bottom plating, longitudinal framing and strength decks, a higher quality steel called *high tensile steel* is adopted which has a tensile strength of 34 to 38 tons per square inch. The beams to decks are formed of angle bulb, i.e. an angle which has a bulb at the lower end, and framing is formed of bars rolled to the shape of a Z or a channel. Pillars supporting decks are formed of tubes, a pillar of seven inches diameter being $\frac{3}{8}$ inch thick. Such pillars are much lighter and stronger than pillars formed of solid material. In the design of the hull structure every effort is made to secure strength with economy of weight, as any weight saved in this portion of the design is available for increasing the fighting or endurance qualities of the ship. An instance of this is seen in the method of specifying plates which are ordered to be so many pounds per square foot and

not by thickness. Variations in thickness are difficult
to detect by measurement but by a simple calculation
of the area the total weight as ordered can be found
and any variation at once detected. Steel material is
liable to rejection if it exceeds the specified weight
and a latitude of $5\,^o/_o$ below this weight is allowed.
Such a condition has an effect on the total weight
of a ship's structure which amounts in the aggregate
to thousands of tons. Angle bars, angle bulbs, and
bars of channel and Z form are ordered by the
weight per lineal foot.

The framing of the ship over a considerable proportion of the length below the lower protective
deck is arranged on the longitudinal system, i.e. the
longitudinal girders are made *continuous* and the
transverse framing is fitted between or *intercostal.*
The central longitudinal girder is termed the vertical
keel and forms the backbone of the ship. Six other
longitudinal girders are fitted on each side of the
centre line as shewn in fig. 2. This system adapts
itself to the formation of a double bottom which is
a most valuable element of safety to a ship. The
longitudinal girders Nos. 2 and 4 together with the
vertical keel are made watertight, thus dividing the
double bottom from side to side into six separate compartments. The longitudinal girder No. 3 each side
is built specially strong and forms a docking keel.
When the ship is docked the blocks are arranged to

come below the vertical keel at the centre line for the whole length and below the docking keels over the central portion of the length where the bottom of the ship is made perfectly flat. The transverse framing over the length of the double bottom is intercostal and is formed of plating connected to longitudinals, and to inner and outer bottom plating, by angle bars. These frames are spaced about four feet apart. The frames are specially strengthened in the neighbourhood of heavy weights as under machinery and barbettes. At convenient intervals the frames are made solid and watertight and with the watertight longitudinals and vertical keel divide the double bottom into a number of separate compartments ; in the ship illustrated in fig. 11 (p. 72) these are 61 in number, 19 of them being arranged for the storage of oil fuel and 5 for the boiler reserve freshwater supply. The remainder, 37 in number, are not utilised for storage purposes but are fitted with watertight manhole covers and are only entered occasionally for inspection, cleaning and painting.

The outer bottom and inner bottom plating are most valuable portions of the structure contributing to the longitudinal strength of the ship and the thickness is determined by the strength calculations already referred to. The plating is doubled below the vertical keel and the docking keels to take the blocks when docking.

At the fore and after ends of the double bottom, where the girth of the ship becomes smaller, some of the longitudinals are dispensed with, being twisted round and incorporated with the fore and aft bulkheads or flats which appear before and abaft the engine and boiler room spaces. These are shewn in figs. 3 and 11.

At the ends of the ship the transverse frames are continuous and formed of Z bars spaced about three feet apart. The longitudinal strength is obtained by means of the numerous bulkheads and flats and fore and aft girders formed of Z bars are fitted as may be necessary.

As shewn in fig. 2 the outer bottom plating is recessed at the lower edge of the armour to form the plating behind armour. The framing here is transverse and of the type shewn in fig. 5, spaced two feet apart in order to give a rigid support to the armour plates. Where the side is continued above armour, as in fig. 5, the framing is formed of Z bars about four feet apart well secured by brackets to the beams and the deck. The beams to the main, upper, and forecastle decks are formed of angle bulbs nine inches deep and these decks are given a 'round-up,' at the middle line so as to determine the flow of water on them to the scuppers at the side. These beams are supported wherever possible by the stiffeners to the cabin and other bulkheads but otherwise by means of pillars. Special fore and aft girders are worked beneath the beams of weather decks in way of gun blast and these are well supported by pillaring or vertical **Z** bars. In the cases shewn in figs. 2, 3 and 5 the upper deck is a protective deck and this is also a strength deck. Such a position for the upper protective deck is admirably adapted for the purpose of effective contribution to the structural strength. Indeed quite apart from protection it would have to be a thick deck to provide sufficient structural strength to the section. Weather decks are covered with teak about three inches thick. This timber is the finest possible for the purpose of durability and for heavy wear and tear. It is somewhat expensive and is getting rather difficult to obtain. The edges of the planks are caulked with oakum to render the deck watertight. The surface of the deck is planed immediately before handing over and the deck is kept in perfect condition by the ship's crew. The teak deck of a warship in commission is really a thing of beauty if not a joy for ever. The use of wood for decks is limited to the weather decks. The steel flats of store rooms, etc. are simply painted to prevent corrosion of the steel. In many places the flats are of chequered or ribbed plating to give a good foothold. In living spaces the steel deck is covered with *corticine*, a strong, thick linoleum, and inside cabins cork carpet is employed. The corticine is laid down in square panels to facilitate renewals and it and the cork carpet are stuck down to the steel deck by a mixture of resin and tallow. A very considerable saving of weight has resulted from the use of these materials in lieu of the wood decks formerly employed.

The use of wood for fittings and furniture is limited as far as possible to lessen the danger of fire especially in action. Thus for instance there is no wood lining in cabins to cover up the steel structure as fitted in passenger ships. The framing in cabins next the ship's side is covered with thin sheet steel to add to the comfort. All cabin furniture of wood, as the washstand, chest of drawers, fronts of bedberths and doors is arranged so that it can be readily removed to be thrown overboard in action. Fittings in store rooms as cupboards, bins, shelves, etc. are constructed of sheet steel as far as possible as also bag-racks, bins for hammock stowage, cap boxes, mess racks, pantry fittings, etc.

Steel castings are employed for places like the stem, stern-post, rudder frames, shaft brackets and shaft tubes. This material can be readily cast to the desired shape and it has a tensile strength of 26 tons to the square inch. Examples of its use are shewn in fig. 16 (p. 113) where the rudder frame, stern casting (*A*), rudder head bearing (*B*) and the crossheads (*C. H.*) are formed of cast steel. The shape of the casting

A gives an idea of the complicated form into which this material can be cast with ribs as desired to obtain the necessary strength with economy of weight.

All watertight divisions or bulkheads are tested both for strength and watertightness by filling compartments with water under pressure. The strength of the bulkheads dividing large compartments, as engine and boiler rooms, is of special importance. These bulkheads, which are of large area and considerable depth, are stiffened by vertical H bars 12 inches deep, well bracketed at the top and bottom. Such a bulkhead 47 feet wide and $25\frac{1}{2}$ feet deep has to withstand a pressure under test of over 600 tons.   In arranging such stiffening economy of weight and economy of space have both to be considered.

Corrosion.   Steel rapidly corrodes if not properly covered with paint. Rust is formed if steel is in contact with moist air containing the gas carbon dioxide (CO_2) and also if immersed in water because of the presence of the gas in the water. Rusting action is much hastened by heat. Besides rust there is corrosion which is caused by galvanic action between steel and other metals (especially copper) if immersed in salt water. The steel becomes badly pitted and shews signs of corrosion. Also steel in the process of its manufacture becomes coated with a hard scale and if this is not removed corrosion will go on as it as well as rust is sufficiently different from steel to

set up a galvanic current with the steel if immersed.
One writer has said that to paint over this mill scale
is like painting a table with the table cloth on.
Special steps are taken to remove mill scale by
immersing the plates in a bath of weak hydrochloric
acid which loosens the scale and it can then be
brushed off with wire brushes.

It is essential therefore to seal hermetically all
the steel structure of a ship by protective paint, and
before this is applied the steel must be thoroughly
clean and dry. For the internal portions red lead is
employed except for confined spaces where oxide of
iron is used as red lead is dangerous to workmen in
such confined places. Three coats are put on except
in store rooms where the first coat is of red lead and
the remaining two coats are white. The upper parts of
the outside of the ship are painted a 'service grey'
and portions not liable to rough usage are 'enamelled.'
The outer bottom under water is usually coated with
three coats of a proprietary protective composition, it
being most desirable to do this work in fine dry
weather.

Fouling. Besides corrosion there is the fouling
of the plating under water to be guarded against.
Marine growths and animals readily attach themselves to the bottom and the additional friction
caused is a serious cause of resistance and loss of
speed. Anti-fouling compositions of various makes

are employed to keep the bottom clean and smooth. The portion of the ship's side 'between wind and water' is especially liable to have the composition rubbed off because of ropes, boats, etc. and the surface in this neighbourhood is covered with a special 'boot topping' composition which is found by experience to be effective in this position. Steel ships are docked at regular intervals to paint the bottom and to renew the anti-fouling composition.

Wherever metal fittings such as sea cocks are attached to the outer bottom we have the conditions favourable to rapid corrosion of the steel by galvanic action and in all such places strips of zinc called 'zinc protectors' are fitted. This zinc protects the steel by itself corroding and whenever the ship is docked these strips are examined and renewed as necessary. They are also fitted in way of the metal propellers and at the entrance of the rudders into the ship (see fig. 16 where marked *Z.P*).

All steel fittings exposed to the weather or damp are *galvanised*, i.e. immersed in a bath of molten zinc after having been thoroughly cleaned. The layer of zinc thus deposited effectively prevents further corrosion.

In order to detect rusting and corrosion at an early stage it is laid down that every accessible part of the structure shall be inspected by the ship's staff once a quarter and any defects discovered are made good, the steel being thoroughly dried and all traces of rust removed before the application of the paint. Hulls are also surveyed by dockyard officers *every four years* in the case of large ships, *every two years* for small vessels and *every year* for destroyers.

CHAPTER III

ARMOUR

A WARSHIP not only has to be capable of inflicting severe blows on the enemy but she has to be capable of withstanding severe blows from the enemy. The armour protection provided for this purpose is used partly for the ship herself to maintain her buoyancy and stability under gun fire and partly for the protection of the guns and gun machinery. The total weight allocated in the design for these purposes will vary in different designs. In the case quoted in Chapter I, 26 % of the total displacement was given to armour protection amounting to about 4250 tons ; this, however, does not include the armour on the revolving gun shields, which is reckoned with the armament.

Successive improvements in the resisting powers of armour have followed the improvements in ordnance, powder and projectiles and it will be of interest to trace briefly the history of the armour question from the time of the *Warrior* (1859) until now, in order to understand the present position.

From 1859 to 1874 the armour which was of wrought iron underwent no change. Improvements in machinery enabled thicker plates to be rolled as time went on. The *Warrior* had armour plates 4½ in. thick, and in successive ships thicker armour was adopted reaching a maximum of 14 in. in the *Devastation* (1869) and 12 in. in the *Inflexible* (1874); in the latter ship two thicknesses of 12 in. were employed to obtain the protection desired.

Compound Armour. In the *Inflexible* however the outer thickness of the turret armour, nine inches thick, was formed of 'compound' or 'steel faced' armour. The object of this was to obtain a hard steel face, by which the projectiles would be broken up, combined by welding with a tough wrought iron back which would prevent the plate from cracking.

The perforation of wrought iron can be represented by a formula devised by Capt. Tresidder and known by his name, viz.:

$$t^2 = \frac{W}{D}\left(\frac{V}{885}\right)^3,$$

where t is the thickness of wrought iron in inches,
V is the striking velocity in feet per second,
D is calibre of shot in inches,
W is weight of shot in lbs. (uncapped).

Any given armour plate which is fired at and perforated can have its thickness compared with that of a wrought iron plate which would be just *perforated* by the shot used, as calculated by the above formula. The ratio of the thickness of the wrought iron plate to the plate tried is known as the *figure of merit*.

For the first compound plates a figure of merit of about $1\frac{1}{4}$ was obtained but successive improvements in manufacture increased this to about 1·7. This armour was employed from the time of the *Inflexible* (1874) up to and including the eight vessels of the *Royal Sovereign* class (1889). In these ships armour of 18 inches in thickness was employed for the belt and 17 inches for the barbettes. Nickel steel armour four inches in thickness was employed for a portion of the side protection above the main 18-inch belt.

Harvey Armour. The development of armour piercing projectiles of forged steel resulted in a further improvement in the quality of armour known as the Harvey process. Harveyed armour had a figure of merit varying from 1·8 to 2·2, and this improvement enabled the belt armour protection to be reduced in thickness to nine inches and increased in area in the next type of ship, the *Majestic* class (1894), of which there were nine built. In the Harvey process a solid steel plate is taken and the proportion of carbon on the face is increased by the cementation process, i.e.

charcoal is placed next the face of the plate (two plates usually being dealt with together, face to face) and the whole is covered in with bricks and run into a gas furnace where it remains for two or three weeks, seven days or so being allowed for cooling. By this means the face to a depth of about $1\frac{1}{2}$ inches has its percentage of carbon increased and is capable of becoming intensely hard when chilled. After cementation the plate is bent to its required shape and all the necessary holes are drilled in the surface. The plate is again heated and the face douched with jets of cold water which gives the hard face required. It is thus a hard faced compound plate without welds and, as seen above, the figure of merit was a distinct improvement on the former compound plates in which the junction between the hard face and the tough back is a mechanical one. The depth of the hard surface being practically a constant for all thicknesses it follows that better relative results are tained with thin than with thick plates, that is to say a 12-inch plate is rather less than twice as efficient as a 6-inch plate. The Harveyed plate however had one defect, viz. that the back was not sufficiently tough to withstand the racking effect of steel projectiles.

Krupp Armour. The Harvey process was soon superseded by the Krupp process of manufacturing armour. The steel employed is of very special quality, of a tensile strength approaching 50 tons per square

inch (ordinary mild steel for shipbuilding has a tensile
strength of 26–30 tons per square inch) and containing
small proportions of nickel, chromium, manganese
and carbon. The surfaces of the two plates face to
face are first carburised, gaseous hydrocarbons being
employed instead of charcoal in some cases. The
surface of the plate is then chilled but the heat of
the plate is carefully graduated from the face to the
back. A special heating furnace is built and the
temperature of the face of the plate raised to a certain
depth sufficiently to allow of the highest degree of
hardness being obtained. The remainder of the plate
is only heated sufficiently to ensure toughness and a
fibrous structure when hardened by water. Hardening is then effected by spraying under pressure the
face *and the back* of the plate simultaneously until
completely cooled. The crystalline structure of the
back of the Harveyed plate is replaced by a tough
fibrous quality which prevents cracking, and by
spreading the resistance to the impact of the projectile over a wider area assists materially in reduction
of penetration.

There are no less than eight heating processes in the
production of a Krupp armour plate, viz. the steel
ingot when cast is not allowed to get thoroughly cool
but has (1) a reheat for pressing, (2) heating for
rolling into a plate, (3) low heat for toughening in
water, (4) heat for cementation, (5) reheat for

hardening in oil, (6) reheat for softening for bending, (7) reheat for annealing after bending (i.e. heated, immersed and allowed to cool slowly), (8) heat for the differential heat treatment.

Such plates are termed Krupp Cemented (K.C.) and have a figure of merit of about 2·3 to 2·7 and in some cases a figure of merit as high as 3 has been obtained. This armour or an equivalent quality has been used in vessels of the Royal Navy since its introduction in 1897, the *Canopus* class being the first ships in which it was employed.

Each manufacturer has presented sample plates to be fired at and if these are considered satisfactory the plates ordered have to be manufactured by a process which is precisely identical with that of the sample plate, resident overseers being appointed for the purpose of inspection. Power is taken under the contracts to withdraw any finished plate desired for the purpose of firing trials and this power is exercised. Cases have occurred in which a plate thus tested has proved unequal in resisting power to the sample and a whole batch of armour has had to be condemned in consequence.

An uncapped projectile of W lbs., D inches diameter and velocity V feet per second will perforate a thickness of K.C. armour of $\dfrac{V}{1500}\sqrt{\dfrac{W}{D}}$ inches approximately, or will penetrate a thickness equal to its own diameter if the striking velocity is about 2500 feet per second (28 miles a minute).

The introduction of 'capped' projectiles has resulted in a considerable increase of penetrating power. The object of the cap, which is fitted on to the hard point of the projectile, is to support the point in a lateral direction to prevent its rupture at the moment of impact.

Owing to the improvement in guns and projectiles and the introduction of the 13·5-in. gun in recent ships, it may be said at the present time that the attack is superior to the defence. The quality of armour has not greatly improved since 1897 while guns have enormously improved and there is also the improvement in projectiles and the introduction of the cap. The following table (taken from the *Navy League Annual*, 1910–11) will illustrate the improvement in artillery since 1897.

	Calibre inches	Weight of projectile in lbs.	Length of bore in feet	Muzzle velocity in feet per second	Muzzle energy in foot tons
Canopus (1897–1900)	12	850	35½	2,367	33,020
King Edward VII (1902–1905)	12	850	40	2,580	39,280
Dreadnought (1905–1906)	12	850	50	3,010	53,400
Orion (1909–1911)	13·5	1250	50	2,821	69,000

(3000 feet per second is at the rate of 34 miles a minute or 2040 miles an hour.)

The following particulars of the performance of a Hadfield's 14-in. 'Heclon' armour piercing shot (capped), given in the 1912 *Naval Annual*, is of interest in this connection.

This shot was fired at a velocity of 1497 feet per second at a 12-in. Krupp cemented armour plate. The projectile passed through the plate and 20 feet of sand butt. No other 12-in. K.C. plate has yet been perforated at this extraordinarily low velocity, which is equivalent to a range of no less than $7\frac{1}{2}$ miles, that is a 12-in. K.C. plate would have been perforated by this Hadfield projectile from a gun placed $7\frac{1}{2}$ miles away.

In regard to recent improvements in the manufacture of armour very little has been made public. The intense hardness of a Krupp cemented plate extends for a moderate depth only. One direction of recent research has been to amalgamate a certain thickness of hard tool steel (which steel has been brought to a high pitch of perfection) with a tough back by a process described below. The following extract from the Engineering Supplement to the *Times* of 29 March 1911, gives a description of an entirely new method of manufacturing armour plates, of a most interesting and surprising character, copper being the agent employed to effect a fusion between the face and the back of the plate.

Simpson Plates.

It was some three years ago that Mr W. S. Simpson, in the course of experiments, discovered by chance that copper and steel, when treated in a certain way, formed a molecular mixture or solid solution of the two metals. After some trials, he found that when two plates of steel, with a section or layer of copper between them, were placed in a mixture of carbon, brown sugar, and water, of the consistency of compressed snow, and the whole mass submitted to a temperature of 2000 degrees Fahrenheit, the copper melted away into the steel and formed a perfect weld. So complete was the union between the two plates that, however treated, they would not part or split at the juncture. Not only was this the case, but the copper, when it entered the steel, increased the tenacity of the latter metal. Samples of such welds were submitted to Professor J. O. Arnold, of the University of Sheffield, who examined them by alternating stress tests and micrographic analysis. Referring to one micrograph he said that it shewed :

'A portion of the thin weld line in which a considerable part of the copper seam has disappeared, owing to the metal dissolving in the steel, to form a solid solution micrographically almost indistinguishable from the steel itself. Had the welding been continued for some time longer, there would be no visible weld line left, but merely a solution of copper in the steel, molecularly continuous with the steel itself, and considerably stronger than the main body of metal, because the copper will have disappeared (visually) after having brought into perfect continuity the molecules of both faces of the steel. The fact that the copper joint is completely soluble in the steel means that to the eye there is no copper visible, it being in a molecular form beyond the range of microscopic vision as an inter-molecular absorption of the copper and steel considerably stronger than the steel itself. The weld, in the ordinary acceptance of the term, no longer exists,

hence securing absolute molecular continuity; it is therefore
obvious that welds made by this process are stronger than the
steel itself.'

Application to Armour-Plating.

It may readily be imagined that such a welding process may
be put to many uses, but for the purposes of the present article
its application to armour plates is especially interesting. As is
well known, by the face-hardening processes which have been in
use for some time past, carburisation of the armour-plate does
not extend to a greater depth than seven-eighths of an inch. A
hard face and a tough back are obtained, but the hard face is
neither as hard nor as thick as could be wished. If, then, it were
possible to take high speed steel, say of double this thickness,
and weld it to a soft steel backing, an immense advantage would
be gained, and this is what can be done with the Simpson weld.
Not only is the high speed steel many times harder than ordinary
steels, but the hardened face of the plate can be made of any
thickness desired. It is understood that trials have been made
already with plates of the usual regulation size in which two
inches of hard steel were welded to four inches of soft, and which,
when attacked by a 6-in. gun, gave entirely satisfactory results.
Larger plates are now being made and will be soon tried, and
although the Simpson armour must be said to be still in an
experimental stage, the results of the trials to be carried out
with it will be certainly looked forward to with more than usual
interest.

In connection with the above the following may
be quoted from the 1912 *Naval Annual*:

'The novelties, from which so much was expected a short time back, have not justified as yet
the promise of the earlier announcements respecting

them. It is rather from improvement in metallurgical processes and by the introduction of new alloys and methods of face hardening, than from any novel systems, that fresh developments in the competition between attack and defence are anticipated.'

Disposition of Armour. The armour provided for the protection of the buoyancy and stability of war vessels is disposed as shewn in fig. 6. The *Dreadnought* and following ships had the armour carried to the main deck and that deck was made a 'protective' deck. Recent vessels commencing with the *Orion* class have the armour carried to the upper deck and that deck is made a 'protective' deck. The belt armour is carried down to the lower protective deck as shewn in figs. 2 and 3, i.e. about five feet below the normal load-line in order to afford protection to the vital parts of the ship when the vessel is rolling or

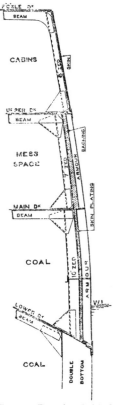

Fig. 5. Framing behind and above armour.

if the ship had a list. The armour has its maximum thickness at the waterline since admission of water there is most prejudicial to stability. A vessel in action however is likely to be 'deep' on account of the coal and oil carried or as the result of damage, and it is important to carry the armour protection well above water, which assists in the preservation of stability at large angles of inclination. In fig. 6 it will be noticed how large a proportion of the broadside area is protected by armour. In the case of the *Dread-*

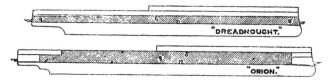

Fig. 6.   Arrangement of Broadside Armour.

nought the armour is carried over the whole length, but in some of the more recent ships this has not been done and the extreme ends are not protected by vertical armour but by strong underwater decks. In this case the citadel of armour is completed by armoured transverse bulkheads. It has been a feature of British designs since the *Majestic* (1895) to protect as large an area of the broadside as possible even if this involved some sacrifice in thickness. This is in contrast to the former French practice in which a very thick narrow waterline belt, extending over the whole length, was preferred.

In this connection it may be mentioned that some of the Russian battleships at the battle of Tshushima built after the French model went into action with their narrow belts completely submerged, being deeply laden, and some were capsized after damage to the side due to gun fire. That is to say they were riddled at and above the waterline and the admission of water reduced the stability to such an extent that they were no longer seaworthy but turned right over and sank.

Gun mountings are protected by circular armoured barbettes as shewn in fig. 3. The thickness of the armour is reduced, as shewn, below the upper protective deck where a shot striking the barbette would also have to penetrate the ship's side or the deck. The rear of the guns and the gun machinery above the barbette are protected by an armoured shield which revolves with the guns and turntable. This also protects the officers and men who are working the guns, etc. The thickness of the shield will usually be the same as that of the fixed barbette, say 11 or 12 inches in a battleship.

Protective Decks. The protection of the ship is completed by means of the protective decks, as shewn in figs. 2 and 3 where the lower and upper decks are such decks. The thickness varies in different designs, a usual thickness being one inch worked in two thicknesses of $\frac{1}{2}$ in.   Hatches in such decks are covered by pieces of plate of the same thickness as the deck and fitted to hinge up.   Where such hatches are necessary for the purpose of escape to men engaged below in action, as e.g. in magazines and shell rooms, they are arranged to open from below as well as from above, and fitted with a balance weight to render the lifting possible from below.   Otherwise such hatches, as those to store rooms, are only arranged to open from above with chain tackles always in position.   These tackles are of the 'non-reversible' type so that the hatch cover cannot run down with its own weight. This is necessary to avoid accidents.

Armour gratings.   There are a number of openings in protective decks that must remain open in action as e.g. those for the funnel uptakes, the ventilation trunks to engine and boiler rooms and to auxiliary machinery compartments.   Such openings are fitted with 'armour bars,' i.e. the opening is divided into a number of sections each of which contains a series of vertical plates about 5 in. deep and $\frac{1}{2}$ in. thick, spaced about $2\frac{1}{2}$ in. apart.   These armour bars give a certain amount of protection while at the same time sufficient area is left to allow of the funnel gases passing up or ventilation being carried on.   Care has to be taken in regard to the security of the bearers in funnel openings to allow of the expansion due to heating.

If securely riveted they would buckle and get distorted when the boilers were lit. The trunks to auxiliary machinery compartments are also provided with watertight sliding shutters at their lower end so that the watertightness of the compartment can be maintained if desired.

Armour gratings placed over machinery have wire mesh gratings worked beneath in order to catch any falling splinters which might pass through the gratings proper. Where escape is necessary from below, as from the engine and boiler rooms, one of the gratings is arranged in three pieces, each of which can be lifted by a man from below. In this case balance weights are not necessary.

Framing behind armour. It is essential that armour should be well supported and this is done by strong framing worked vertically about two feet apart, with fore and aft supporting girders. The vertical frames are well supported at the decks by means of large brackets. The plating behind armour, which is really the outside of the ship's structure, is about ¾ inch in thickness and teak backing is worked between it and the armour. This backing was formerly of great thickness to form a cushion, e.g. in the *Admiral* class it was 15 inches thick, but in the latest ships the backing is simply a pad to form an even bed shaped to the exact shape of the back of the armour. The minimum thickness is about two inches behind

the thickest armour, increased as necessary behind
the thinner armour to form a flush surface on the
outside (see fig. 5). This backing is secured to the
skin plating behind armour by galvanised steel bolts
screwed through the plating and secured at the
inside by nuts with washer and hempen grommet in
order to ensure watertightness. A specimen frame
behind armour is given in fig. 5 by which it is seen
that the framing is formed by 10-inch and 7-inch
Z bars with brackets, flanged at the edges for stiffness,
at the decks and beams. A fore and aft girder runs
along connected to each frame. Besides supporting
the armour, these frames, of course, perform their
function of providing the transverse framing of the
ship herself.

Armour bolts. The armour plates have to be
effectively connected to the structure of the ship and
this is done by armour bolts which screw into the
back of the plate and bear on the skin plating behind
armour with washer and hempen grommet as indicated in fig. 7. With the wrought iron armour
formerly used the bolts were taken right through the
thickness with a conical head let into the surface, but
with hard faced armour this would be undesirable as
the holes in the face would facilitate cracking when
the plate was struck. On this account hard faced
plates are secured at the back as indicated above.
It will be noticed that in the bolt shewn in fig. 7 the

shank is made slightly less in diameter than the
screwed portion of the bolt at the bottom of the
thread. This being so the shank is the weakest portion of the bolt and would tend to stretch on the
rebound after a blow on the armour. If the bolt at
the thread were the weakest part there would be the
likelihood of the bolt breaking there and the armour
falling off the ship. The annular space between the

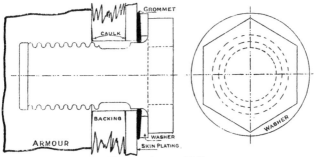

Fig. 7. Armour Bolt.

bolt and the backing is filled with a water-excluding
mixture. A considerable number of bolts are used
for each plate, so that if the plate were badly cracked
in action the portions would remain on the ship, being
thus at any rate partially effective as protection.

The size and shape of armour plates are given to
the manufacturers by means of wooden moulds. This
information is prepared in the mould loft of the shipyard where the vessel is 'laid off,' as it is not possible

to wait until the ship is far enough advanced to get the shapes. This work is done by very skilful mechanics who rarely make a mistake. It may easily be understood how expensive a mistake would be in regard to these moulds. The manufacturer of the armour plates has to bend them, often in two directions, to form the curved surface of the ship and this work is performed with great skill. There is an element of judgment required as the plate is bent while hot and the alteration of shape on cooling has to be allowed for. All the arrangements which require holes in the face of the armour have to be worked out in the shipyard drawing office at an early stage. Such are the holes for scupper drain pipes, sanitary pipes, etc., and connections for the torpedo net defence fittings. The information is supplied to the manufacturer who cuts the holes and drills and forms the thread in small holes before the final hardening takes place. The surface cannot be drilled after hardening unless the material is softened locally and this of course cannot be allowed as it would destroy the hardness of the armour. The final finish of the edges of armour plates is done by a process of grinding so that they will fit accurately to one another. With belt armour it is usual to allow for 'closers,' i.e. plates for which the final neat sizes are only given when the remainder of the armour is bolted on board. This is necessary because there may be slight

differences between the vessel as laid off and as built.
The plates for circular barbettes are erected complete
at the maker's works and they can be made so accurately as to fit hard up against the circular ring of
plating behind armour erected on the ship, and
'closers' in this case are not allowed for.   In order
to obtain accuracy in this work standard gauges are
used by the armour makers and the shipbuilders.

CHAPTER IV

ARMAMENT

WORKERS in three distinct fields of science have
combined in order to produce the marvellous perfection of modern artillery.   There is the chemist
who by research and experiment has obtained a
propellant which is smokeless and which is of comparatively slow combustion, thereby pushing the
projectile the whole time it is in the bore of the gun.
The influence of the propellant is seen in the great
length of modern guns, and the enormous velocities
that are obtained.   The slow combustion has caused
moderate pressures to be obtained in the gun,
and this has had great influence on design.   The
old powders were too rapid in their action, the
pressure rapidly falling off as the projectile travelled

down the bore and there was continual search after
a 'slow burning' powder in order to avoid excessive
pressures in the breech of the gun.   There is also the
influence of the metallurgist who by research and
experiment in the qualities of steel has enabled steel
of absolute reliability and enormous strength to be
produced for the manufacture of guns and projectiles.
And there is the mechanical engineer who has produced the machinery for operating the guns.

We shall deal with these divisions briefly in
order, viz. :—Propellants, .Guns, Gun mountings and
machinery.

Propellants.   The cordite used in the British
Navy is composed of a compound of guncotton,
nitro-glycerine, and a small percentage of mineral
jelly.

Nitro-glycerine is a powerful explosive formed by
the action of nitric acid on glycerine with sulphuric
acid to absorb moisture.   Guncotton is a highly
nitrated cotton.   The mixture of these two substances
will not take place as the latter is not soluble in the
former, but this is rendered possible by the use of
acetone as a solvent.   Mineral jelly is added, this
having a waterproofing effect, and if a very small
percentage of certain alkalis are added the stability
of the mixture is considerably improved. The resulting
mass can be pressed or rolled and cut into any suitable
form desired.   The name 'cordite' comes from the

cords into which the gelatinised mass was squeezed
out of a cylinder with a hole in one end. Cordite is
an eminently safe explosive and a piece of it can be
burnt freely like a candle without any explosion. It
was a very striking discovery that the union of two
such explosives as nitro-glycerine and guncotton,
neither of which could be used in guns because of
the instantaneous nature of their explosion, should
have resulted in *cordite,* a material that has such
valuable properties. Not the least valuable of these
qualities is that it is smokeless. The primary importance of the stability of the propellant used on
board ship was brought home in a most forcible
manner by the loss of the French battleship *Liberté* in
1911. The Commission which enquired into this
disaster reported that it was not due to malevolence
or fire but due to the inflammation of a cartridge
of service powder in a magazine, and stringent
regulations were issued as to the manufacture of the
powder which appears to have been carried out in
a faulty manner, departures from accepted processes
having been allowed. The powder used in the French
Navy is of the type called '*nitro-cellulose*' powder,
which is also used in the Navies of Russia and the
United States. The *nitro-glycerine* powder is used in
the Navies of Great Britain, Germany, Austria, Italy
and Japan. There is every reason to believe that
the qualities of British cordite as regards stability

are perfectly satisfactory, but cooling machinery is fitted on all British warships to maintain an even temperature of about 70 degrees in the magazines. Variation of temperature is found to affect the ballistics of cordite.

As already stated cordite gives a comparatively moderate pressure on being fired and the pressure falls off but slowly as the projectile travels along the gun. The intensity of this pressure is about 18 tons to the square inch, falling to 10 tons at the muzzle. This has been the cause of the increase of the length of guns because an effective pressure is maintained behind the projectile for a comparatively long period.

Guns. The guns fitted in a modern warship are formed of high grade steel of absolutely uniform quality, high tensile strength and great ductility. The metallurgy of steel has had great attention paid to it in recent years, especially in Sheffield, and results are now obtained which a short time ago would have been considered impossible. Lieutenant Sir Trevor Dawson gave the following table in a lecture in 1909 which illustrates the enormous advance in the material now available compared with that in 1864. At that time steel was brittle and unreliable and quite unsuited to the construction of a gun.

Tests for material for an 1864 *gun and a* 1909 *gun*
(tons per square inch).

	1864		1909	
	Breaking Stress	Elastic Limit	Breaking Stress	Elastic Limit
Wrought iron coils	23	12	—	—
Wrought iron hoops	10	—	—	—
Ordinary steel for guns	—	—	34 to 44	21
Nickel steel for guns	—	—	45 to 55	30
Wire for guns	—	—	90 to 110	—

These qualities of steel may be compared with ordinary mild steel used universally for ordinary ship construction which has a tensile breaking strength of about 30 tons per square inch.

The inner tube of a gun is 'rifled,' i.e. it has spiral grooves formed in it for its whole length. On the rear of the projectile is a soft copper ' driving band ' and this is forced into the grooves and the projectile is thus given a rotation as it passes down the gun. This spin of the projectile is of great importance as it brings into play a gyroscopic effect which keeps the projectile point forward. Without this effect it would turn round at right angles to its line of flight. It is interesting to note that a projectile maintains the same angle to the horizontal throughout its flight as it had when leaving the muzzle of the gun.

The barrel of the gun is placed on a lathe and revolved and steel wire ribbon is wound round. This

ribbon, which is about $\frac{1}{4}$ inch wide and $\frac{1}{16}$ inch thick,
has a tensile strength of 100 tons to the square inch.
The ribbon is wound on from end to end at varying
degrees of tension and upwards of 120 miles of ribbon
is used in the construction of a 12-inch gun. The
gun is completed by shrinking on further steel tubes
and the whole gun is machined externally in an
enormous lathe.   At the rear of the gun is the breech
block, one of the most perfect pieces of mechanism
it is possible to conceive and which has been
rightly described as a 'triumph of mechanics.' The
interrupted threads on the breech block engage with
corresponding threads in the breech of the gun and
one twelfth of a turn is sufficient to secure the breech
block in the breech of the gun. The swinging in of
the breech block and the locking are performed by
hydraulic power in about four seconds and about the
same time is occupied in unlocking and opening.
Hand power can be used in case of a breakdown,
when the operations take about six to seven seconds.
    The gun is fired by means of an electric current
which fires a small portion of a high explosive which
in turn ignites the powder charge. By this means
a round can be fired at the instant desired which is
important when the gun platform (the ship) is a
moving one and all the guns can be fired simultaneously from some central position by the mere
pressing of a button. Ships have thus fired a

simultaneous broadside without the slightest damage to the structure.

The fumes from a charge were formerly found a serious drawback; when the breech was opened these came into the turret and time had to be spent in washing out the gun by a water spray. The roof of the gun shield had to be perforated to allow the fumes to escape. The introduction of an air blast has been an immense improvement. This comes into operation automatically as the breech commences to swing back and the chamber and barrel of the gun are cleared before the breech block is fully opened.

Gun machinery. The operation of the guns and the transport of the projectiles and ammunition require a large number of special auxiliary machines. All recent ships (except the *Invincible*) use hydraulic power for these purposes. Electric power was fitted in the battle cruiser *Invincible*. The hydraulic system has special advantages. It is reliable and adaptable and for running the gun back to its firing position, for working the rammer into the breech of the gun and for operating the lifting cage containing the projectiles and ammunition, the direct motion of the hydraulic ram is more easily applicable than the rotary motion of the electric motor. The hydraulic system is also exceedingly safe, and damage can be at once located and easily repaired. Several hydraulic pumping engines worked by steam power are installed, each in a separate compartment, giving a hydraulic pressure of about 1000 lbs. to the square inch and pressure pipes are taken round the ship under protection with branches to the various barbettes. Exhaust pipes take the used-up water back to the hydraulic tanks.

By reference to fig. 3 it will be seen that the revolving turntable has a trunk extending down to the shell room and this revolves with the mounting. The hoist of the projectile and ammunition for each side takes place up to the top of this trunk and thence another cage takes them to the rear of the gun. This 'stepped' supply is conducive to rapid firing as when the upper cage is going up to the gun the lower cages are going down for a fresh supply. The hoists on each side for each gun work independently. The projectile is lifted from the shell bin by a grab and, suspended from overhead rails, worked by hydraulic power, is transferred to a bogie which revolves round the trunk. It can then be rolled into the cage which is hoisted to the top of the trunk when the shell is tilted on to a waiting tray. Similarly the four quarter charges are put into the hoist at the magazine level and lifted to the top of the trunk whence they are tilted on to a tray. Both shell and powder hoists are fitted with automatic safety catches for use if the lifting wires break. The shell and charges are transferred to the upper cage

by means of hydraulic rams. This cage is then
hoisted to the rear of the gun and is so arranged to
stop at the gun at any angle of elevation or depression
of the latter. The contents are transferred to the
gun, projectile first, by means of a chain rammer
worked hydraulically. This rammer is formed by a
series of links, jointed like a two-foot rule, which are
rigid in one direction but which will fold up in the
other. After the rammer has pushed the charge into
the gun it returns and forms a flexible chain which is
stowed under the loading arm on the gun. The
various operations described above have automatic
checks which prevent any portion of the gear operating
before its time. Thus the rams to push the projectile
and powder charges on to the upper cage cannot
operate until the upper cage is in the bottom position
ready to be loaded. Also the chain rammer cannot
start unless the cage is level with the breech and this
cage cannot descend until the rammer is completely
withdrawn. It is seen from the above that the guns
can be loaded at any angle of training and any degree
of elevation and the supply arrangements enable the
rate of fire from a twin 12-inch mounting to be from
three to four rounds per minute with projectiles
weighing 850 lbs. each.

The turntable, trunk, shield and guns are revolved
by means of hydraulic motors controlled from the
sighting positions. These are in duplicate and are each

powerful enough to rotate the mounting supposing the ship has a list of ten degrees. The turntable revolves on a special roller path with roller bearings so arranged as to reduce the friction to a minimum.

The recoil of the gun is absorbed by two cylinders placed at the rear of the gun slide and attached thereto. The principle is to force the liquid from one side of a piston to the other through a small orifice in the piston and the energy of the recoil is thus absorbed. The gun is 'run in' or 'run out' by hydraulic cylinders and elevated and depressed by similar means.

(A detailed account of all the above gun machinery is given in Mr McKechnie's paper read before the Institution of Naval Architects in 1907.)

The following remarks concerning accuracy of fire taken from Sir Trevor Dawson's lecture already referred to may be quoted. 'The result of the development in the engineering side of artillery, in the optical accessories and in great measure in the enthusiasm, training and skill of the gunners, is a marked improvement in the battle practice of the fleet. The target which is towed *at an unknown speed, course and range,* is only 90 feet long by 30 feet high, less than one-fifth the length of a battleship. One of the latest cruisers put in 18 hits out of 32 rounds from 12-inch guns, and another 15 hits out of 18 rounds. Taking all guns in all British ships in commission, the percentage of hits to rounds fired in 1908 was 56 % under the new and more severe firing conditions. It may be taken that efficiency in gun practice has more than trebled in ten years.'

A set of five twin 12-inch mountings with hydraulic engines, guns, and gun shields, as installed in the *Hercules* costs about £510,000 or about £102,000 per mounting. A set of five twin $13\frac{1}{2}$-inch mountings etc. as installed in the *Orion* costs about £550,000 including guns or about £110,000 per mounting. This of course does not include the cost of the fixed barbette armour fitted for the protection of the mounting. The muzzle energy of a 12-inch gun is 53,400 foot tons and of a $13\frac{1}{2}$-inch gun is 69,000 foot tons or in the ratio of 1 to 1·3. The cost of the guns and mounting is in the ratio 1 to 1·08 so that the power has been increased 30 % at an increase of cost of only 8 %. This comparison does not consider the extra cost of the larger ship which would be necessary to carry a given number of the heavier guns and mountings.

Disposition of the main armament. Fig. 8 shews the disposition of the main armament on British ships since the *Dreadnought*. In that ship the main armament consisted of 10 12-inch guns in five mountings and this same number of guns has been adhered to in all British battleship designs since, but the calibre has been increased to $13\frac{1}{2}''$ in the *Orion* and later classes. There have been three arrangements adopted and the special feature of the later ships has been the superposed turret and the fact that all the guns can fire on either broadside. The following table shews what guns can fire ahead, astern, and on the broadside in typical ships designed on these three systems.

		Dreadnought		Neptune		Orion
Ahead	...	6	...	6	...	4
Astern	...	6	...	8	...	4
Broadside	...	8	...	10	...	10

The dispositions of the armament of these three vessels are shewn in fig. 8. In the *Dreadnought* the forward turret is above the forecastle deck level and all the others above the upper deck level. The turrets on the broadside are in the same transverse section. Seven vessels were built with this disposition, viz.:— *Dreadnought, Bellerophon, Superb, Temeraire, St Vincent, Collingwood,* and *Vanguard,* and the displacement increased from 17,900 tons to 19,250 tons.

In the *Neptune, Hercules,* and *Colossus* the foremost turret is above the forecastle deck level. The broadside turrets are above the upper deck level and arranged *en échelon* so that each can fire on the opposite broadside. The two after turrets are brought together and the foremost one is raised so that the guns can fire over the aftermost turret.

In the *Orion* and later ships the new $13\frac{1}{2}$-inch gun was adopted and the turrets are all arranged on the centre line. The second turret from forward and the second turret from aft are arranged so that the

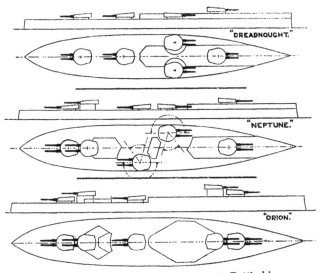

Fig. 8. Arrangement of main armament: Battleships.

guns can fire over the foremost and aftermost turrets respectively. It will be noticed that the average height of the guns above water in this system is very much greater than in the *Dreadnought* and this has been the main cause of the considerable increase of breadth which is necessary for stability purposes. The comparative dimensions of the three vessels shewn in fig. 8 are given in the following table.

	Dreadnought	*Neptune*	*Orion*
Length	490 ft.	510 ft.	545 ft.
Breadth	82 ft.	85 ft.	88 ft. 6 in.
Draught	26 ft. 6 in.	27 ft.	27 ft. 6 in.
Displacement	17,900 tons	19,900 tons	22,500 tons
Main Armament	10 12-inch	10 12-inch	10 13½-inch

The disposition of the main armament of the battle cruisers has been arranged in three different ways. First in the *Invincible, Indomitable* and *Inflexible* the 8 12-in. guns were arranged as in fig. 9, three pairs on the forecastle level and one pair aft on the upper deck.   The midship guns are arranged *en échelon* but close together so that although the starboard guns could fire on the port side and *vice versa,* in an emergency, the two turrets could hardly be worked simultaneously on the same broadside on account of blast.   The next ship the *Indefatigable* had these guns more widely separated as in the *Neptune* (fig. 8). In the *Lion* and *Princess Royal* the 13½-in. gun was adopted and the eight guns are all arranged on the centre line as shewn in fig. 9.   The two forward turrets are close together, the after guns being able to fire over the forward turret.   Very considerable angles of training are obtained with all the guns and four fire ahead, two astern, and the whole eight on either broadside.

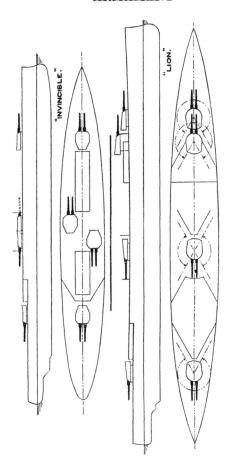

Fig. 9.   Arrangement of main armament : Battle-cruisers.

The comparative dimensions of these three classes are as follows:

	Inflexible	Indefatigable	Lion
Length	530 ft.	555 ft.	660 ft.
Breadth	78½ ft.	80 ft.	88½ ft.
Draught	26 ft.	26½ ft.	28 ft.
Displacement	17,250 tons	18,750 tons	26,350 tons
Main Armament	8 12-in.	8 12-in.	8 13½-in.
Designed Speed	25 knots	25 knots	28 knots

In all the above ships each pair of guns is arranged to train over as large an arc of training as possible, but as will readily be seen by reference to fig. 8 it would be possible in some cases for one gun to fire into the gun of a neighbouring mounting as e.g. if the *Neptune's* forward guns were trained on the starboard beam and the starboard guns were trained forward. To avoid any damage due to this, 'portable stops' are arranged on each mounting which prevent the mounting being trained beyond the limits of absolute safety. When necessary these stops are released and when the relative positions of training of two mountings are such that damage would ensue if a gun were fired a buzzer starts and continues sounding in the endangering turret so long as the endangered turret continues in the danger zone. These are fitted on Kilroy's patent system of electrical danger signals. Similar signals are provided for turrets which are close together where, although clear for horizontal fire at all angles of training,

damage is possible with the guns at different degrees
of elevation and depression.

Minor armament. This armament is provided
for use against torpedo craft. In the *Dreadnought*
the 12-pounder gun, 24 in number, was adopted, the
following being an extract from the 1906 Admiralty
memorandum dealing with the design.

'In view of the potentialities of modern torpedo craft and
considering especially the chances of torpedo attack towards the
end of an action, it is considered necessary to separate the anti-
torpedo boat guns as widely as possible from one another, so that
the whole of them shall not be disabled by one or two heavy
shells. This consideration led the Committee to recommend a
numerous and widely distributed armament of 12-pounder Q.F.
guns of a new design and greater power than those hitherto
carried for use against torpedo craft.'

Subsequent ships have carried a larger gun, the
four inch, the numbers provided being 16 for the three
ships of the *Bellerophon* class, 20 for the three ships
of the *St Vincent* class and 16 for the *Neptune,
Hercules* and *Colossus* and the *Orion* class.

Torpedo armament. A torpedo is an under-water
weapon which on being pushed out of the tube in the
ship is self propelled at a high speed by means of
compressed air and it has a great radius of action.
Automatic appliances are embodied in the torpedo
for regulating the depth and for keeping the torpedo on the desired bearing. In the head of the
torpedo there is a bursting charge of guncotton which

will explode on striking an obstacle. In ships other than small cruisers, destroyers, and torpedo boats, torpedoes are always fired from submerged tubes. The number of such tubes varies in different designs, in some ships three and in others five, one of these being right at the stern and the others on the broadside. The size of torpedo adopted for many years was the 18-in. but in the *Hercules* and *Colossus* and subsequent ships the 21-in. torpedo has been arranged for.

The following table from the 1912 *Naval Annual* gives the present ranges and speeds of the torpedoes constructed by Messrs Whitehead & Co. of Fiume.

Diameter of Torpedo	Speeds in knots				Explosive charge
	1000 yds.	3000 yds.	6000 yds.	8000 yds.	
18 in.	$42\frac{1}{2}$	—	27	—	209 lbs.
21 in.	—	41	—	27	330 lbs.

CHAPTER V

EQUIPMENT

In the vessel referred to in Chapter I, $4\,\%$ of the total displacement or 655 tons was devoted to the equipment consisting of fresh water, provisions, officers' stores, officers, men and effects, anchors and cables, masts, rigging, etc., boats, warrant officers' stores and net defence. A few remarks will be made as to these various items.

Fresh water. This is stowed in storage tanks and arrangements made for its distribution by electrical pumps in duplicate to the daily supply tanks high up in the ship whence it is led to the various washplaces, galleys and issue tanks. Considerable care is necessary to prevent the waste of fresh water as water is a very precious commodity on board ship. Provision is made for the production of fresh water by distillers placed in the engine rooms but water is taken on board from waterboats whenever possible. It may be stated that only senior officers have a water supply tap in their apartments. Ordinary cabins are simply provided with a washbasin with water can and drain can below.

Provisions. The amount allowed depends on the

complement of officers and men and the period for
which the ship is stored.

Officers' stores. This includes the food and other
stores carried for the use of the Admiral, Captain,
Ward Room and Gun Room and also the stores
carried by the Paymaster for the use of the crew, as
cloth, soap, etc.

Officers, men and effects. This depends on the
number of the complement, each officer and man with
his effects being taken to weigh $2\frac{1}{2}$ cwt.

Anchors and cables. There are usually three large
anchors carried, two bower and one 'sheet' with
corresponding amounts of cable stowed in cable
lockers. A smaller 42 cwt. anchor is supplied, near
the stern if possible, for use in holding the stern of
the ship in a stream. Steel wire rope is provided for
this anchor and also for use when being towed by
another ship.

Masts, rigging, etc. This will include also the
derricks provided for boat lifting and for coaling
purposes.

Boats. There are usually two steam boats supplied,
56 ft. or 50 ft. long and in the case of an Admiral's
ship a steam barge. Other boats include a 42-foot
sailing launch and several life cutters. Two of these
are hung in davits ready for immediate lowering at
sea in the event of a man falling overboard. The
heavy boats are lifted out of their crutches and

lowered at the side of the ship to the water by means
of a strong derrick operated either by hydraulic or
electrical machinery which is controlled from above
the weather deck. All this gear is tested by suspending
weights about twice the weight that has to be dealt
with in practice. The pulling boats are largely used
for exercising the men of the ship and a good deal of
sailing work is also done with them.

Warrant officers' Stores include the stores placed
under the charge of the Carpenter and the Boatswain
and those under the charge of the Gunner and Torpedo
Gunner which are not directly connected with the
armament of the ship.

Net Defence. This consists of the fittings supplied
to protect the ship against torpedo attack. A steel
crinoline is hung down over a large proportion of the
length at the ends of long steel booms which stand
out from the side of the ship a short distance above
the water level. Arrangements are made to pull the
booms aft to stow against the ship's side and at
the same time to furl the nets, so that when the head
of the boom comes in at the level of the net shelf the
nets come in too and roll over on to the net shelf
which runs along the edge of the weather deck. (See
frontispiece for booms and net.)

CHAPTER VI

STABILITY AND ROLLING

Stability. One of the most essential features to be obtained in a warship is the provision of sufficient stability, not only for ordinary seagoing purposes but to ensure that the ship shall remain stable and sea-worthy after sustaining a reasonable amount of damage in action. The conditions as regards stability are special as compared with merchant vessels because of the great weights carried high up for armour and armament which involve a relatively high position of the centre of gravity. The initial stability or stability near to the upright (10 to 15 degrees from the vertical) is governed by what is known as the *metacentric height* or the distance between the centre of gravity and the metacentre. The former point depends on the vertical distribution of the weights forming the aggregate weight of the ship at any particular time, and will vary in position according to the magnitude and distribution of these weights. The latter point depends on the form of the ship and especially on the shape of the ship at the waterplane at which she is floating at any particular time. A broad waterplane will lead to a high metacentre and a narrow waterplane to a low metacentre. This may be illustrated by the

familiar case of a square log floating half immersed,
which will always be found to float corner downwards.
If placed with the top and bottom surfaces parallel
to the water level, the breadth of the waterplane is
the breadth of the log and this leads to a position of
the metacentre *below the centre of gravity* and the
log is unstable and cannot float in this position.   If
however it is placed with one corner of the section
downwards the breadth of the waterplane is the
diagonal of the square section and this leads to a
metacentre above the centre of gravity and the log
will float stably in this position.   For stability the
metacentre must be above the centre of gravity and
the distance it is above, or the metacentric height, is
a measure of the stiffness of the ship.   A small metacentric height means that the vessel is easily inclined
by external forces and a large metacentric height
means that she will be difficult to incline.   A small
metacentric height, however, leads to a long period of
oscillation and a ship of small metacentric height is
less likely to meet a sea which synchronises with
(i.e. suits) her period than a ship of greater metacentric height and shorter period of oscillation.   Thus
a small metacentric height leads to easy motions in
a seaway and passenger liners with a reputation for
steadiness at sea have metacentric heights varying
from a few inches to two feet, the actual amount in
a given ship varying on different voyages with the

distribution of the cargo and on the same voyage with the consumption of coal. Such small values of the metacentric height are not permissible in war-vessels because it is necessary to allow for the reduction of the area of the waterplane, by the admission of water due to gun fire, which pulls down the metacentre. It was the sudden reduction of the waterplane in the *Victoria* due to the admission of water to the superstructure through open doors that rendered the ship unstable and caused her to capsize as she was settling down by the bow after the collision with the *Camperdown*. (See the author's *Warships* for an account of this catastrophe.)

War-vessels are provided with broadside armour over a large proportion of their length and in some cases over the whole length. In the former case the ship at the fore and after ends of the broadside armour is provided with transverse armour bulkheads and the protection of the vital parts below water is obtained by thick armoured decks (see figs. 6 and 11). The armour is thickest at the waterline and the ship in this neighbourhood is minutely subdivided by longitudinal and transverse bulkheads (see fig. 11 at middle deck) to localise the inflow of water and so limit the reduction in waterplane area, supposing the armour was pierced. Where the extreme ends are not protected by vertical armour the loss of stability due to riddling of the unarmoured ends would be relatively

small, and the metacentric height provided is amply
sufficient to give proper stability with the unarmoured
ends completely gutted.   The amount of metacentric
height obtained in battleships varies of course with
the load on board especially in regard to coal and oil,
about five feet being probably an average value.
Battle-cruisers have a rather less metacentric height.
The metacentric height given to war-vessels is thus
seen to be greater than would be arranged for if easy
motions in a seaway only were desired, but the continued stability of the ship after damage in action is
the essential requirement to be met.

Besides initial stability, the stability at large
angles has to be carefully considered.   Many large
merchant vessels with a small metacentric height can
be inclined from the upright to 75 degrees or beyond
before they would become unstable and capsize,
assuming the side and deck watertight.   This amount
of stability at large angles is obtained by a relatively
low centre of gravity and a high freeboard.   Although
war-vessels have a freeboard greater than is required
by the load line regulations for merchant vessels, their
centre of gravity is relatively very high because of
the great weights of armour and armament high up
in the ship to which merchant vessels have no equivalent.   A high centre of gravity is prejudicial to
a large *range of stability* (or angle from the upright
beyond which the vessel will capsize) and it is found

usually in battleships that this range is in the neighbourhood of 60 degrees, supposing the sides and deck
of the vessel are intact and watertight. The stability
at angles beyond the upright must also be considered
supposing the unarmoured portions of the broadside
are riddled in action. All British war-vessels starting
with the *Majestic* (1895) have their armour carried
well above water, to the main deck in some ships and
to the upper deck in others, this armour being reduced
in thickness above the waterline belt. A large metacentric height has considerable influence in lengthening
out the range of stability and this is one of the main
reasons for the larger metacentric heights given to
battleships as compared with battle-cruisers. In the
latter ships the freeboard for a considerable proportion of the length extends to the forecastle deck
(see fig. 9) and this lengthens out the range as
compared with battleships in which the forecastle is of
much more limited extent (see fig. 8). It will be
noticed from the dimensions of the *Hercules* and the
Indefatigable given in Chapter I that the former has
considerably greater breadth which results in a much
higher metacentre.

The righting moment (or effort the ship makes to
return to the upright) at any given angle of inclination
can be calculated and its variation with the inclination
can be shewn graphically by what is known as a *curve
of stability*, measurements along the base-line being

angles of heel, and lengths of ordinates righting
moments. Fig. 10 shews curves of stability for a battleship and a battle-cruiser, the range of the former being
60 degrees and of the latter 75 degrees. The greater
slope of the curve of the battleship at the start is
due to the greater metacentric height or stiffness and
the way in which the curve lengthens out for the
battle-cruiser is due to the greater freeboard. The

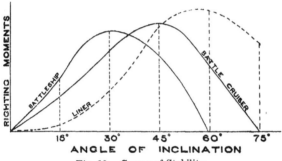

Fig. 10. Curves of Stability.

area enclosed between such a curve of stability and
the base-line is a measure of the work that must be
done on the ship by external forces to capsize her
and the larger this area the greater is the reserve of
stability possessed by the ship against damage of
the side in action. The dotted curve is a curve
of stability similarly constructed for an Atlantic
Liner in which the initial stiffness is small but the

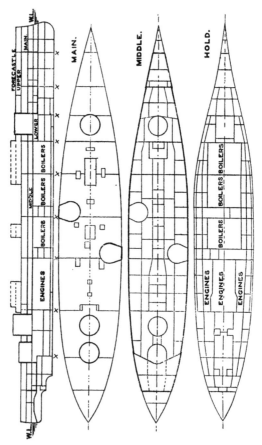

Fig. 11. Watertight subdivision of a Battleship.

range of stability is very great, well beyond 75 degrees.
The author is indebted to W. J. Luke, Esq., of Messrs
John Brown & Co., Ltd., Clydebank, for permission to
include this curve of stability.

Rolling. It may be stated generally that to
produce a ship that shall be easy in a seaway she
should be given a small metacentric height and this
is always done in large passenger liners, usual values
being from a few inches to two feet. This gives them
a long period of roll and the likelihood of such a ship
meeting with a sea which suits her period, causing
heavy rolling motions, is very remote. With larger
metacentric heights, however, the period of roll is
shorter and a sea which suits the ship's period is
more likely to be met with and this will result in
heavy rolling. The essential conditions of warship
design have been seen to necessitate somewhat large
metacentric heights and under certain circumstances
heavy rolling may result and this is more likely to
happen if oil fuel is carried in the double bottoms by
which the stiffness is considerably increased. It is
very desirable that warships should have easy motions
in a seaway and not roll much because of the fact of
the ship being a gun platform. Rolling is minimised
by fitting *bilge keels* or fins projecting out from the
bilge of the ship each side over the midship portion
of the length (see fig. 2). At the fullest portion of
the ship these have usually been made of small depth

to enable the vessel to pass through the entrances of certain dry docks. If larger entrances are available the bilge keels may be made of considerable depth all along and they are then most effective in reducing rolling motions as they have their full depth at the portion of the ship where they act at the greatest leverage about the centre of oscillation. Bilge keels act in reducing rolling motions not only by the direct resistance offered to their passage through the water but also in lengthening the period of the ship and tending to upset the synchronism (or timing) of the sea with the period of roll of the ship. Anti-rolling tanks have been adopted in some large liners on a system devised by Herr Frahm. These depend on the transference of water from one side of the ship to the other and the system appears to have been successful in reducing rolling. Such tanks were tried in the *Edinburgh* and other ships many years ago, but their adoption is a matter of considerable difficulty on account of the space involved, but it is understood that such tanks are to be tried in some ships of the Royal Navy now building.

The following extract from a paper read by Sir W. H. White, K.C.B., before the American Society of Naval Architects in November 1911 may be quoted in connection with the stability and rolling of war-vessels.

Modern capital ships—with heavy-gun armaments and great weights of armour placed high above water—must be endowed with considerable initial stability in order to possess a reasonable range of stability under ordinary peace conditions. Hence they have been made very broad in relation to their draughts of water, even after allowance is made for the increase in draught (deep laden) as compared with their predecessors.

Their moderate range of stability (when intact) gives to these ships a less margin of safety when injured in action by gun-fire; and, in many instances, the distribution of the armour is of a character that adds to this risk, because it leaves without protection large areas of the sides which are reckoned as water-excluding in calculations of stability, although they can be rapidly riddled and destroyed by shell-fire with high explosives. In later designs the areas which are armour-protected have been considerably increased; but the addition to the weight of armour placed high above the waterline has necessarily been accompanied by a rise in the position of the centre of gravity of the ships. The breadths have been still further increased in proportion to the draughts, and even then the range of stability in the intact condition is very moderate.

Modern warships, notwithstanding their greater size, are not, and cannot be, as steady gun-platforms as their predecessors, because their great initial stability causes them to have relatively short periods of oscillation. As a consequence, they are more likely to be set rolling by ordinary conditions at sea. Experience has shewn them to be less steady in a seaway than ships of earlier date and smaller size. Some of these ships are so large in relation to the available dry-docks that their bilge keels are either very shallow or non-existent at the midship portions of the length; and as a consequence the check upon rolling is greatly reduced.

CHAPTER VII

ENGINES, BOILERS AND AUXILIARY MACHINERY

THE propelling machinery of the *Dreadnought* and subsequent ships is of the Parsons turbine type, and the power has been divided into four units driving four propellers, as compared with twin screws in previous battleships and cruisers. The way had been prepared for this new departure in the comparative trials of the third class cruisers *Topaze* and *Amethyst*, which were sister ships except that the former had reciprocating engines driving twin screws and the latter had turbine engines driving triple screws. These ships were laid down in 1903 and were of the following dimensions, viz.: 360 ft. long, 40 ft. beam, 14 ft. 6 in. draught and 3000 tons displacement and for reciprocating machinery were designed for $21\frac{3}{4}$ knots, with 9800 I.H.P. A careful series of comparative trials was carried out when the *Topaze* obtained 22·1 knots with 9933 I.H.P. and the *Amethyst* obtained 23·63 knots with an equivalent power of 14,200. From a similar installation of boilers (although not of the same type) the turbine ship gave 43 % more power.

The following statement was officially issued in

1906 by the Admiralty regarding the adoption of the turbine machinery in the *Dreadnought*.

'The question of the best type of propelling machinery to be fitted was most thoroughly considered. While recognising that the steam turbine system of propulsion has at present some disadvantages, yet it was determined to adopt it because of the saving in weight and reduction in working parts and reduced liability to breakdown; its smooth working, ease of manipulation, saving in coal consumption at high powers and hence boiler room space and saving in engine room complement; also because of the increased protection which is provided for with this system, due to the engines being lower in the ship: advantages which much more than counterbalance the disadvantages. There was no difficulty in arriving at a decision to adopt turbine propulsion from the point of view of sea-going speed only. The point that chiefly occupied the Committee was the question of providing sufficient stopping and turning power for purposes of quick and easy manœuvring (the turbine does not reverse like reciprocating machinery). Trials were carried out between the sister vessels *Eden* and *Waveney* (torpedo boat destroyers) and the *Amethyst* and *Sapphire* (third class cruisers), one of each class fitted with reciprocating and the other with turbine engines; experiments were also carried out at the

Admiralty Experimental Works at Haslar and it was considered that all requirements promise to be fully met by the adoption of suitable turbine machinery and that the manœuvring capabilities of the ship, when in company with a fleet or when working in narrow waters will be quite satisfactory. The necessary stopping and astern power will be obtained by astern turbines on each of the four shafts. These astern turbines will be arranged in series, one high and one low pressure astern turbine on each side of the ship, and in this way the steam will be more economically used when going astern and a proportionately greater astern power obtained than in the *Eden* and *Amethyst.*'

As an illustration of the economical use of the steam by the turbine it may be mentioned that if the *Dreadnought* had been fitted with reciprocating machinery to give equal power, she would have required 22 boilers instead of the 18 actually fitted.

Turbine machinery runs at considerably higher revolutions than reciprocating machinery. The pistons of the latter require to be brought to rest at the beginning and end of each stroke and with large engines this imposes a limit on the speed of revolution, the maximum accepted for such engines in the British Navy being 120 revolutions per minute. With turbine machinery there are no reciprocating parts and the engine is perfectly balanced as regards rotation and for turbine installations on land very high rates of revolution are usual. On board ship, however, the turbine has to work in conjunction with the propeller and a propeller working at high revolutions is not efficient as a propelling instrument. A compromise has to be accepted, viz. the turbine is run slower than would conduce to its maximum efficiency while the propeller is run faster than would be desirable, the aim being to obtain the most efficient combination. Thus the *Dreadnought's* turbines were designed for 320 revolutions per minute which is much faster than previous reciprocating machinery of equal power. Propellers for turbines have to be of comparatively small diameter in order to avoid an excessive velocity at the tip. The magnitude of this velocity may be seen when it is stated that the tips of the blades of a propeller of 10 feet diameter at 320 revolutions a minute have a velocity of 170 feet per second.

In order to obtain the blade area required, on a limited diameter the area of the blades has to be proportionately very large. The relation of the blade area to the disc area in turbine practice goes up to about 60 %, about 35 to 40 % being the usual previous practice. The blade area is determined by the amount of thrust each propeller has to exert and it is found that if the blade area is insufficient the phenomenon called *cavitation* sets in, that is to say, the water will not follow up at the back of the blade and cavities of water vapour are formed which cause a serious loss of thrust.

The cycle of operations in a set of turbine machinery is as follows: The steam from the boilers is taken into two main steam pipes to the port and starboard engine rooms.   The steam is taken to the high pressure turbine where its pressure gradually drops and thence to the low pressure turbine whence it proceeds to the *condenser*.   The condenser is a nest of tubes through which a continuous flow of sea water takes place, the circulating pumps being of the centrifugal type.   The steam coming into contact with the outside of these tubes becomes condensed and is pumped away by means of an air pump to the hot well from which it is pumped back again to the boilers. With turbines the usual arrangement is to fit a high pressure turbine on the outer shaft and a low pressure turbine on the inner shaft on each side of the ship. In the earlier turbine vessels of the Royal Navy a 'cruising' turbine was fitted on each centre shaft in order to economise coal at low powers, but in recent arrangements this has not been done and the weight and space saved have been utilised to increase the efficiency of the main turbines.   Sir Chas. Parsons has stated that the result of this modification has been to increase the efficiency of the main turbines at full power, and this increased efficiency is felt down to

about half power. Below half power, however, the
efficiency of the turbines is somewhat less than that
of the earlier combinations with cruising turbines.
In the earlier ships the turbine machinery was
arranged in two independent and isolated compartments. With the increase of the power and the
larger engine rooms required it has been considered
desirable to subdivide the total space further.   In
some ships three engine rooms are provided, as in
fig. 11, with two intact watertight longitudinal bulkheads between.   In others four engine rooms are
provided, formed by a longitudinal middle line bulkhead and a transverse bulkhead.   In this case the
forward engine rooms contain the turbines and the
after rooms the condensers, circulating pumps, and
steering machinery. The transverse bulkhead has of
course to be pierced by the large 'eduction' pipes
which convey the steam from the low pressure turbines
to the condensers.

The turbine is not reversible and an 'astern'
turbine is fitted on each shaft, sometimes independent and sometimes in the same casing as the ahead
turbines.   In order to go astern it is necessary to
operate the steam valves to shut off the supply to the
high pressure ahead and open the supply to the high
pressure astern turbine. Special arrangements are
made by which this operation can be rapidly performed.

Boilers.   The boilers fitted in ships of the British

Navy are of the watertube type. In the former
cylindrical boiler the water was in bulk and sur-
rounded the furnaces and the tubes which conveyed
the products of combustion to the uptube and funnel.
The watertube boiler on the other hand has the water
inside tubes with the fire outside. By this means the
total weight of the boiler and water is considerably
reduced and what is most important from a military
point of view, steam can be raised much more quickly
than with a cylindrical boiler. Watertube boilers
require very careful design and handling, because of
the small amount of water in them, in order to avoid
the water level falling, but experience has shewn them
to be admirably adapted to the conditions obtaining
on war-vessels. There are two types of watertube
boilers fitted, viz. the *Yarrow* and the *Babcock and
Willcox*. It would be beyond the limits of this work
to attempt to describe in any detail the features of
these boilers. It may be stated that arrangements
are fitted on the boilers for the use of oil fuel which
is stored in some of the spaces between the inner and
outer bottoms.

Auxiliary machinery. A very large number of
auxiliary engines are provided on board the modern
warship. These are usually manufactured by firms
who make a speciality of such engines. The hydraulic
pumping engines are referred to in the chapter on
armament, and the water under pressure provided by

these engines works a great variety of hydraulic engines for the operation of the heavy guns and mountings, and in some ships the hydraulic boat-hoisting machinery. Two steam steering engines are employed in separate main engine rooms for operating the steering shaft which runs aft to the steering gear. Either can be coupled with this shaft as desired. This duplication is provided in case one engine breaks down or one engine room is flooded.

Electrical machinery is largely adopted for various purposes; the ship is lighted throughout by electric lamps and current is also necessary for the searchlights. The electric current is obtained from several electric generating machines placed in widely separated iso-lated compartments. These have usually been driven by reciprocating engines but in some ships a turbine-driven dynamo has been fitted, running at about 3000 revolutions per minute. The standard voltage now adopted is 220 and this has to be transformed to a lower voltage for the operation of the searchlights. Transformers are also provided to produce a 15-volt current for telephones and firing circuits. Among the many electric motors fitted may be mentioned the motors for ventilating fans, some of which are of large size, 40 in. and 50 in. diameter, for ventilating the engine rooms, the usual size for the ordinary ship ventilation being $12\frac{1}{2}$ in. diameter.

The after capstan for warping is driven by an

electric motor and in some ships the hoists for operating the heavy boats are electrically driven. The 50-ton centrifugal pumps for fire and drainage purposes and the fresh water pumps for lifting the fresh water from the storage tanks to the daily supply tanks are driven by electric motors. There are also electric winches used for coaling, torpedo net defence work, lifting the light boats, and for general purposes. Electric lifts are provided, one to each engine and boiler room, to facilitate the passage of the officers for supervision purposes, there being no access between these compartments by watertight doors. The magazine cooling and refrigerating machines are all electrically driven and the sounding machine is also provided with an electric motor. The electrical energy for lighting and power is taken right round the ship by means of a *ring main* cable, from which branches are taken as necessary and the circuits are so arranged that a duplicate supply of current is taken to important positions to minimise the possibility of the light failing altogether. A very complete system of telephones is adopted; these are direct between important positions and a telephone exchange is also provided for general use.

A very powerful steam-driven capstan engine is provided forward under protection for operating the cable holders and capstan on the forecastle deck, for working the anchors and cables.

Besides the above there are the auxiliary engines in connection with the main machinery. The ventilating fans for supplying air under pressure to the boilers are steam driven. Fire and bilge pumps are provided in each engine room and boiler room for dealing with water in those compartments and for providing a supply of salt water for fire purposes as an alternative to the 50-ton motor-driven centrifugal pumps. Feed pumps are fitted for supplying the boilers with fresh water. The centrifugal pumps for circulating water through the condensers and the air pumps have already been referred to and one of each of these is provided for each main and auxiliary condenser. Oil pumps are provided for the oil fuel. Air-compressing machinery is provided, with reservoirs for storing the compressed air for the purpose of the torpedo installation and the air blast to the large guns. Distilling machinery for evaporating sea water is provided to obtain a supply of fresh water for boiler or drinking purposes, with the necessary pumps to circulate the water gained. Ashes are discharged from the ship by ash ejectors which pierce the side of the ship above water, one for each boiler compartment. They are worked by water under pressure provided by the fire and bilge pumps.

The above list of auxiliary machinery is by no means exhaustive, but it gives some indication of the large amount of special machinery that has to be provided and looked after, quite apart from the propelling apparatus proper of the ship.

CHAPTER VIII

VENTILATION AND PUMPING

THE ventilation arrangements of a warship have to be very carefully considered, not only for the purpose of obtaining an efficient supply of fresh air and exhaust of foul air, but also to ensure that the ventilation pipes do not interfere with the watertight divisions of the vessel. The problem has been much simplified by the adoption of electrically driven fans. The system may be divided into the following groups, viz. :

Ventilation of Boiler Rooms.
Ventilation of Engine Rooms.
Ventilation of Coal Bunkers.
Ventilation of Auxiliary Machinery rooms, as for dynamos.
Ventilation of Store Rooms.
Ventilation of Magazines.
Ventilation of Cabins and Living Spaces.

Ventilation of Boiler Rooms. Boiler rooms are provided with steam-driven fans which deliver air under pressure for the supply of air to the boiler furnaces. These fans incidentally provide sufficient ventilation to the boiler rooms, the exhaust taking place through the boilers and up the funnels. The supply is taken down large watertight trunks which are continued up to the weather deck, armour gratings being provided at the level of protective decks. These trunks also provide the means of access to the boiler rooms. Electrically driven fans are not suitable for this work on account of the high temperatures in the upper part of the boiler rooms in which the fans are placed.

Ventilation of Engine Rooms. In this case large electrically driven fans are adopted, fans being provided both for supply and exhaust with trunks leading to the weather deck. Provision is made for natural supply and exhaust when it is not desired to use the fans.

Ventilation of Coal Bunkers. Special attention is necessary for this ventilation because of the gas that comes from coal. This gas forms an explosive mixture with air and if allowed to accumulate is likely to lead to explosions and injury to the men entering the bunkers. The gas is given off more freely from the coal if the temperature rises and the temperature of coal bunkers is ascertained at frequent intervals, thermometer tubes being provided in all bunkers for this purpose. The ventilation of coal bunkers is carried out by natural means, i.e. fans are not employed. At the top of each bunker a supply

pipe and an exhaust pipe are fitted. The exhaust pipe
is led into an adjacent funnel casing and up to a
position above the weather deck and this being hot
will cause the air in the pipe to rise. Thus a current
of air is induced over the top of the bunker and a
supply of fresh air comes down the supply pipe, which
is led down from a position above the weather deck,
usually down a cool ventilating trunk. When the
boilers are working under air pressure it is necessary
to close these supply and exhaust pipes to prevent the
escape of air from the boiler rooms. The upper ends
are provided with locked louvres and it is laid down
that these louvres must not be closed without special
orders.

Ventilation of auxiliary machinery spaces. These
spaces will be occupied in action as the dynamos,
hydraulic machinery, etc. will be necessarily then in
use and the spaces containing the steam machinery
are hot and need a continual supply of fresh air.
Electrically driven fans are employed sometimes on
the exhaust system and sometimes on the supply
system. In the former the air is drawn out of the
top of the compartment and discharged above the
weather deck, the air-supply being drawn down
the access and escape trunk. With this system
efficient ventilation is obtained without draughts as
the fresh air coming down the large trunk is moving
at a low velocity. With the supply system a fan

draws down fresh air which is distributed round the compartment by means of trunks and is exhausted up the escape trunk. In each case the fan is fitted behind armour protection. The supply and exhaust trunks are made watertight and armour bars are fitted at 'protective' decks, and in addition sliding watertight shutters are fitted at the lower part of the ventilating trunks which can be operated from a position above the water-level and which would be closed in the event of the compartment being abandoned.

Ventilation of Store Rooms, etc. For this ventilation the sections into which the length of the ship is divided are each provided with several electrically driven fans and the watertight bulkheads marked by a cross in fig. 11 are not allowed to be pierced for ventilation purposes below the main deck. These fans draw air by means of trunks from the weather deck and each discharges into a series of pipes, usually six in number, which are led to the various compartments which the fan serves. Proper provision is made to secure the watertightness of the ship by means of watertight pipes and watertight valves so that if one watertight compartment is flooded the water cannot find its way into any other compartment. The exhaust to the various compartments is provided through the doors and hatches and the use of the ventilation arrangements is limited to the minimum amount necessary, for a certain period during the day. At other times and especially during the night or during an action all such ventilation valves, doors and hatches are closed and must not be opened without the permission of the Officer of the Watch. Many compartments are not directly ventilated but are ventilated as required by means of a hose connected to an adjacent ventilation trunk and led through the hatch or doorway. This is done to limit as far as possible the piercing of bulkheads and flats by ventilation pipes.

Ventilation of Magazines. The ventilation of magazines containing cordite is incorporated with the cooling arrangements which are designed to keep the temperature of the magazines at about 70 degrees Fahrenheit. A brine cooling system is installed, the brine being brought to a low temperature by means of refrigerating machinery. The brine is circulated through a cooling tank which consists of a nest of tubes through which the brine passes. The air in the magazine is exhausted by means of an electrically driven fan and passes over the tubes in the tank ; the air thus cooled then passes again into the magazine. The continual circulation of the air in this way causes a lowering of the temperature in the magazine. The fan can draw fresh air from a trunk leading from the weather deck and an exhaust is provided for getting rid of the foul air. Every care is taken to prevent magazines being heated from adjacent compartments, well ventilated air spaces being provided where necessary, and lagging of a non-conducting material is fitted on bulkheads to prevent the conduction of heat.

Ventilation of Cabins and Living Spaces. This ventilation is incorporated with a heating system. The fan draws air from a mushroom ventilator on the weather deck and discharges it through a 'heater' which consists of a nest of copper tubes through which steam is passed. The heated air is then distributed by means of trunks round the messing and sleeping spaces and into the cabins. The heater is fitted with a 'bye-pass' so that the air can be delivered straight through without coming into contact with the steam pipes.

Exhaust fans. For special compartments in which the air gets very foul a natural exhaust is found insufficient and besides the forced supply of fresh air, a forced exhaust is provided by means of an electrically driven fan. Such places are the washplaces, seamen's waterclosets and urinals. The compartments containing cooling machinery are provided with exhaust fans as here dangerous gases are likely to accumulate.

PUMPING AND FLOODING

Arrangements are necessary to provide a supply of water for fire or wash-deck purposes, to provide means of pumping out the bilges and double bottom

compartments into which water is likely to find its
way, to provide means for flooding compartments for
the purpose of correcting heel or trim due to damage
and also for flooding magazines and shell rooms in
the event of fire taking place on board. Arrangements
are also necessary to allow magazines to be flooded
when the vessel is in dry dock with ammunition on
board. The vessel is divided into sections and the
bulkheads marked with a cross in fig. 11 are unpierced
by any piping, each section being an independent unit.
Powerful electrically driven centrifugal pumps are
installed, each capable of delivering 50 tons an hour
at a high pressure. The suction is taken from the
sea and a rising main for the delivery is taken up
vertically from each pump. These rising mains are
connected and branches taken where required with
fittings, on to which hoses can be readily attached
for fire or wash-deck purposes. The steam 'fire and
bilge' pumps in the engine and boiler rooms are also
used for a similar purpose, emergency fittings being
provided for use in case the pipes above the protective
decks are damaged in action. The centrifugal
pumps are also arranged with suctions leading to
various bilge and double bottom compartments with
a discharge overboard, and arrangements are made so
that these compartments may be flooded if desired.
Special arrangements are provided for flooding the
bunkers and wing compartments abreast the engine

and boiler rooms in order to provide a means of correcting heel due to damage. It is quite conceivable that a ship might receive damage on one side in action which would give her a heavy list, possibly sufficient to bring the lower edge of the armour on the opposite side out of the water. Because of this and to enable the guns to be fought it would be necessary to endeavour to bring the vessel as nearly upright as possible and this can be effected by the flooding arrangements described.

Provision for flooding magazines is of supreme importance in order that the ammunition may be drowned in the event of fire. One of the finest vessels of the French Navy, the *Liberté*, was totally destroyed in the year 1911, with great loss of life, owing to the ammunition in a magazine exploding. This disaster was due to unstable powder, but it illustrates the frightful catastrophe that might result if a vessel caught fire in the neighbourhood of a magazine. Every magazine is provided with means of flooding direct from the sea. In order to prevent accidental flooding two locked valves are employed, one at the seacock and one at the magazine, both of which must be simultaneously open to allow flooding to take place. Each of these valves can be operated from the main deck or from a compartment adjacent to the valve, this alternative being provided in case one position was untenable owing to fire or other causes.

Whenever a vessel having ammunition on board is placed in dry dock, fittings are fastened at the openings to the seacocks with hose connections on to which hoses can be secured and these hoses would be led to the water supply of the dockyard.

CHAPTER IX

WATERTIGHT SUBDIVISION

THE watertight subdivision of a war-vessel has to be considered not only from the point of view of navigation risks but from the point of view of the damage likely to be received in time of war. On account of the latter the subdivision arranged for is far greater than can be given to merchant vessels. In the latter there is usually an inner skin forming a double bottom over a certain proportion of the length at the lower part of the ship and also watertight transverse bulkheads and in some ships watertight longitudinal bulkheads at the side of the ship abreast the boiler rooms, between which and the ship's side coal is stowed. With the watertight bulkheads in a merchant vessel it has been the practice up to the present to provide watertight doors in the bulkheads separating the engine and boiler rooms from one another for the purpose of effective supervision,

and similar doors are necessary on watertight longitudinal bulkheads to coal bunkers to get the coal
out to the stokeholds. On the ships belonging to the
great steamship lines carrying passengers all these
doors are arranged with a closing arrangement,
usually hydraulic, which can be operated from the
bridge and which also acts automatically in the event
of water rising on either side of the bulkhead. It
was in 1906, at the time of the *Dreadnought* design,
that the extreme step was taken in ships of the
British Navy of abolishing altogether watertight
doors between machinery spaces : the following is
an extract from the Admiralty memorandum.

'Special attention has been given to safeguarding the ship
from destruction by underwater explosion. All the main transverse
bulkheads below the main deck (which will be nine feet above the
waterline) are unpierced except for the purpose of leading pipes
or wires conveying power. Lifts and other special arrangements
are provided to give access to the various compartments.'

Practically the only important bulkheads below
water which have watertight doors are those dividing
the boiler rooms from the coal bunkers and these
doors are necessary for getting coal out of the bunkers.
They are vertical sliding doors which can be operated
from a position well above water as well as at the
door itself. The arrangements referred to above
for closing watertight doors by hydraulic or other
means have not been considered desirable in the

British Navy. The following extract from a Parliamentary Paper concerning the loss of H.M.S. *Victoria* is appended as giving the then Admiralty view as to these appliances.

'In doors and scuttles there are the risks of the automatic appliances failing to act, or of solid materials being carried into openings by a rush of water and preventing doors from closing properly. These considerations have led to the retention of existing fittings the design of which provides that when properly closed and secured, doors and hatchway covers shall be as strong as the neighbouring partitions and watertight under considerable pressure. There is no difficulty in making automatic appliances. It is a question of which plan secures the maximum of safety under the working conditions of the Royal Navy. With large numbers of disciplined men, familiar with the fittings and constantly drilled in their use, it is possible to close and properly secure all the doors etc. in a battleship in three to four minutes or possibly a less time for ships after long periods in commission.'

There are four sets of watertight divisions in a large warship, viz.:

(i) A watertight inner skin about $3\frac{1}{2}$ feet from the outer bottom which extends transversely to the protective deck (as shewn in fig. 2), and fore and aft over a large proportion of the length, in one case 70 %. This inner skin is valuable in lessening the possibility of water obtaining access to large compartments if the outer bottom were damaged owing to a sunken rock or other similar causes. Cases are on record where the existence of such an inner bottom has saved a ship from foundering.   The spaces between the inner and outer bottom are well subdivided by means of watertight transverse divisions at convenient intervals and by the longitudinal framing being made watertight as indicated in fig. 2.

Many of these spaces are arranged to carry oil fuel and some are arranged to carry fresh water for the reserve feed supply of the boilers.   Although the inner bottom as such is not extended right to the fore and after ends there is a virtual inner bottom at the ends formed by means of the watertight flats and platforms.

(ii)   *Watertight Transverse Bulkheads* extending well above water.   A number of these are regarded as main bulkheads and must not be pierced either for doors or for ventilation or drainage purposes. The system adopted is to divide the length of the ship into a number of sections bounded by the main bulkheads (in the case illustrated in fig. 11, ten in number marked ×) and these sections are made separate and distinct as regards pumping, flooding and ventilation, and no doors are allowed to be cut in them up to the level of the main deck.   There is thus no communication possible fore and aft below that deck and the safety of the ship in the event of damage is immensely increased thereby.   Each of the boiler rooms and engine rooms has an electric lift which rises to the main deck level and allows quick

passage to be made between these compartments
by the officer in charge for supervision purposes.
There is also telephone communication. If voice
pipes are provided between compartments below
water they are led well above water up and down
again. The transverse bulkheads between the main
and upper and the upper and forecastle decks have
hinged doors for the necessary fore and aft access.
These doors would all be closed in action.

(iii) *Watertight Longitudinal Bulkheads.* The
principal of these are the longitudinal coal bunker
bulkheads and the engine room bulkheads. The
former below the protective deck necessarily have
doors to get the coal out to the boilers but they can
be worked and closed from above as well as below.
The latter bulkheads are necessary to isolate the
machinery so that in the event of one engine room
being flooded the other set of engines and steering
machinery would be available. These latter bulkheads
are treated in the same way as the main transverse
bulkheads above mentioned, i.e. they are not pierced
for any purpose whatever except as may be necessary
for power or electric leads. It will be noticed (see figs.
2 and 11) that two intact fore and aft bulkheads are
worked between the main and middle deck over the
length of the machinery spaces on either side of the
openings for engine and boiler ventilators and funnel
casings. At the ends of the ship, before and abaft the

engine and boiler rooms, there are a large number
of fore and aft bulkheads to form the boundaries of
magazines, store rooms, etc. These are in most cases
unpierced for doorways, access to the compartments
being obtained by watertight hatchway covers. When
doorways are necessary they are hinged and watertight
and they are only opened for access to the store rooms
and for ventilation and kept closed at night and when
the vessel is in action.

(iv) *Watertight Decks and Flats.* All the decks
and flats are made watertight and hatchway covers
are capable of being closed and made watertight.

Compartments which would be occupied in action
like the Submerged Torpedo Room, the Dynamo
Rooms and the Hydraulic Machinery Compartments
are each provided with a watertight escape trunk
leading well above water and this trunk at its lower
end is provided with a watertight sliding shutter
which can be operated either from above or below.

Fig. 11 shews the watertight subdivision in a
battleship. The forward bulkheads are carried well
above water and the foremost compartments of all
are not allowed to be used for stowage purposes, the
bulkheads and flats being simply provided with bolted
manhole covers sufficiently large to allow a man to
get through for periodical inspection during painting.
These forward bulkheads are of the greatest importance as it is necessary to ensure them remaining

intact and efficient after collision. The ship shewn
in fig. 11 had upwards of 335 separate and distinct
watertight compartments below the upper deck, 61 of
which were double bottom compartments. Such
minute subdivision is of course impossible in merchant
vessels on account of the spaces required for cargo
and passengers but the extra expense and complica-
tion of working has to be accepted in war-vessels to
provide for the possibility of damage in time of action.
It will be noticed that between the middle and main
decks, i.e. in the neighbourhood of the waterline, the
bulkheads are numerous. This is in order to localise
the damage and prevent undue reduction of the waterplane area, this being prejudicial to the stability.
None of these bulkheads either longitudinal or
transverse are pierced by doors. It will also be
noticed that transverse watertight bulkheads are
fitted at intervals between the main, upper, and
forecastle decks. The ventilating trunks to engine
and boiler rooms are carried above the upper deck
and made watertight, and watertight doors are fitted
in them above the main deck level for access and
escape.

Watertesting. During the construction of the
ship each of the compartments below the main deck,
with the exceptions named below, are tested by
actually filling with water, the head of water which
determines the pressure being taken well above

the water level at which the vessel floats. This test
not only determines the watertightness of the surrounding bulkheads but also tests their strength.
The bulkheads right forward are very strongly built
and tested to a severe pressure test to ensure them
remaining intact supposing the vessel were down by
the bow. One engine room and one boiler room is
filled with water to the required height and the
surrounding bulkheads are carefully examined for
watertightness and for deflection. Other similar
large compartments are not filled, but the surrounding
bulkheads where not tested by adjacent compartments
are tested for watertightness by means of a hose
delivering water at high velocity. The magnitude
of the operation of filling an engine or a boiler room
will be understood when it is stated that in one case
no less than 1850 tons of water were required for this
purpose. A flat 30 feet below water of area say
16 feet square would have to stand a pressure of
220 tons supposing a compartment above or below
were bilged, so that substantial work is seen to be
absolutely necessary to avoid rupture or distortion
sufficiently serious to destroy the watertightness.

All these tests are carried out with the watertight
doors and hatches in position as they would be in the
completed ship and after the watertesting is satis-
factorily finished the work of fitting up the store
rooms, installing the electrical equipment, etc. is

proceeded with. In arranging fittings and leads for these purposes the watertight portions of the structure are avoided wherever possible, but piercing of the bulkheads cannot be entirely avoided and there is always the possibility of small holes being made and not used : the presence of such holes would destroy the desired watertightness. A most rigid and careful examination has to be made of all watertight bulkheads, etc. shortly before the completion of the ship to see that no holes are left, and to ensure this being carefully carried out by the builders a searching air pressure test is insisted upon. It is thus seen that in the practice of the British Admiralty all is done that is possible to ensure that every compartment called watertight is really watertight and will remain so under any pressure that it may reasonably be expected to be subjected to in service.

It goes without saying that the continued efficiency of the ship as regards watertightness depends on the supervision and repair as necessary of the fittings on hatches and doors, and regulations have been issued to the ships of the fleet that once a year the Admiral Commanding is to require that a selected compartment is filled with water in each ship.

CHAPTER X

STEERING AND TURNING

WAR-VESSELS require to have a very much greater turning power than is necessary in vessels of the mercantile marine as they have to manœuvre in company with other ships in a fleet. The rudders must be relatively large in area, they must be capable of being put hard over quickly, and the shape of the stern must be so designed as to facilitate rapid turning. Steam steering gear is always employed, and this is fitted in duplicate in case of breakdown, with capability for steering by manual power as a further alternative. As a comparison it may be stated that the *Lusitania*, the fastest Atlantic liner, turns in a circle of diameter four times her length or 3000 feet and in the *Dreadnought* battleship the circle is 2·7 times her length or 1320 feet.

It is an essential condition that the rudders and steering apparatus of a warship should be well under water and under protection and this is the reason of the shapes given to the stern (see figs. 12, 13 and 14). Up to the time of the *Dreadnought* (1905) all large vessels had a single rudder and with the length of battleships in vogue up to that date a single rudder was found to be all that was necessary for manœuvring purposes. Many battleships were cut up at the stern to improve their turning qualities and a 'balanced' rudder fitted, that for the *Lord Nelson*, shewn in fig. 12, being an example. A 'balanced' rudder is one having a portion of its area before the axis, the centre of the water pressure being then close to the axis of rotation. This not only means that the steering engine and gear may be less powerful than for a ship having a rudder unbalanced (i.e. hinged at the fore edge) but the steering gear can take the rudder over to its extreme angle very quickly and steering by manual power is also rendered easier of accomplishment. The battleships immediately prior to the *Dreadnought* were those of the *King Edward VII* and the *Lord Nelson* classes, of 425 feet and 410 feet length respectively. In the *Dreadnought* the increased length of 490 feet was necessary to arrange effectively the 12-inch guns forming the main armament. Now an increase of length, other things being equal, means a loss of turning power, and if the *Dreadnought* had been fitted with a single rudder her turning circle would have had a diameter in the neighbourhood of four times her length and it would have been difficult for her to have manœuvred in company with other battleships with very much smaller turning circles. In view of this the stern was so arranged as to have completely underhung twin rudders in the same transverse section. These

rudders were immediately abaft the inner propellers
and this means that the 'steerage' (or power of
turning before the ship has started any motion ahead)
is greatly increased, because directly the engines
start there is a stream of water against the rudders.
The large rudder area obtained by having two
rudders combined with the cut up at the stern
resulted in an extraordinary facility for turning, as
may be seen in the following comparison with the
Minotaur, an armoured cruiser having a single rudder
with the stern cut up like the *Lord Nelson* (fig. 12).

		Tactical Diameter		
	length in feet	Yards	In terms of ship's length	Area of middle line plane ÷ rudder area
Minotaur	490	600	3·7	48·4
Dreadnought	490	440	2·7	37·5

The tactical diameter is the distance between the
course of the ship when the rudder is put hard over
and the course when the ship has turned through a
half circle.   The shape of the rudder and the stern
adopted in some recent battleships and battle-cruisers
is given in figs. 13 and 14.   The rudder is 'balanced'
rendering it comparatively easy to turn, as the centre
of the water pressure is never far from the axis of
turning.   There is at full speed and full helm angle a
great 'bending moment' on the rudder head where
it enters the ship, as there is no bottom support.
The rudder of course is made the subject of careful

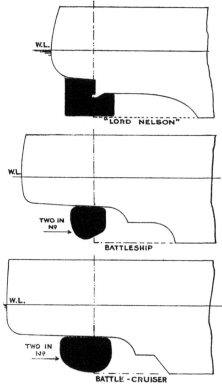

Figs. 12, 13, and 14.

design and calculation to ensure that it shall be of
ample strength in all its parts. The pressure per
square foot on a rudder varies as the *square of the
speed*, i.e. if we double the speed we have pressures
four times as great to provide for. Comparing
rudders of battleships of 21 knots with battle cruisers
of 28 knots the ratio of pressure per square foot is
$(21)^2$ to $(28)^2$ or one to 1·8 nearly, and comparing a
battleship of 21 knots with a destroyer of 36 knots
the ratio is nearly one to three. This is one illustration of the great increase of the forces that have to be
dealt with as speeds increase.

The rudder heads are kept below the protective
deck at the stern and are thus well below water and
under protection. The rudders and steering gear are
such vital portions of a ship's equipment that this is
absolutely necessary to preserve them from damage
due to gun fire in action. (It is of interest to note in
this connection that the Cunard liners *Lusitania* and
Mauretania, which are liable to be taken over by the
Admiralty in time of war for use as cruisers, have the
rudder and steering gear well under water, and thus
less liable to damage. For the sake of appearance,
however, the sterns above water appear the same as
those of ordinary merchant vessels. See fig. 15.)

Each of the rudders is keyed to a crosshead worked
by a set of screw steering gear. This steering gear
is formed by a right and left handed screw thread

which has two nuts, so that as the screw is made to
revolve the nuts come together or recede according
to the direction of rotation of the screw. These
nuts are joined by connecting rods to the crosshead
above mentioned and so by rotating the screw shaft
the rudder is turned as desired, the usual maximum
angle of helm being 35 degrees from the middle line

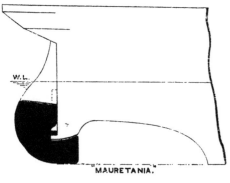

Fig. 15.

in both directions or 70 degrees in all. The two sets
of screw gear are connected across the ship by wheel
gearing so that they move together simultaneously.
Other types of steering gear have been adopted but
the screw gear has now been fitted in all large ships
of the Royal Navy for some years. Although this

gear is wasteful as regards power it has the very
great advantage of being very compact, a matter of
considerable importance at the fine ending of the
stern. It can also be made 'non-reversible,' i.e.
blows on the rudder do not pass beyond the screw
gear, the gear being then termed 'self-holding.' This
is specially valuable when steering by manual power.
The shaft operating the two sets of steering gear runs
from the steering gear compartment direct to the
engine room where it engages with one of the two
steam steering engines in the two engine rooms. The
power operating the gear is thus duplicated and either
steering engine can be clutched to the steering shaft
as desired. This would be valuable in the event of a
steering engine breaking down or one engine room
being flooded. The main steering shaft can be dis-
connected from the steering gear and gearing from
the hand wheels connected up. The hand wheels
cannot operate the steering gear at full speed nor at
the full helm angle, and are of course much slower than
the steam gear, but for ordinary steering purposes the
hand gear is a valuable stand-by in case the steam
engines were both out of action. If the steering gear
and rudders on both sides were disabled the only
alternative then for steering the ship is by means of
the propellers. Cases have been known in which
owing to the loss of a rudder a twin-screw vessel has
been effectively steered by altering the revolutions

of the engines. The same thing would be possible in the quadruple screwed ships now built.

The steering engines are operated from the bridge and conning tower by small hand wheels. The usual method of communication is by means of a 'telemotor' system. A small piston is worked by the steering wheel and small pipes filled with liquid are led from it to the steering engine, an identical motion being obtained at the engine to that at the wheel. These telemotor pipes are carried down through an armoured communication tube and then aft under protection to the steering engines in the Engine Rooms. Compasses are provided at all the steering positions.

A systematic series of turning trials is carried out for every war-vessel and recorded for the information of the Commanding Officer. These consist of the following, viz. : (a) full speed, (b) 12 knots and (c) one engine ahead and one astern 12 knots, rudder assisting; all are taken to port and starboard respectively. From the observations taken the paths of the ships can be plotted and the following information tabulated, viz. :

The *Advance*, i.e. the distance from the position when the helm is put over until the ship turns through 90 degrees or at right angles to her original course.

The *Tactical Diameter*, i.e. the distance between the centre line of the ship when the helm is put over until she has turned through 180°, i.e. with the course reversed.

On first putting the rudders hard over there is a tendency for the vessel to heel inwards but this only lasts a short while as when the vessel is on her circular path the centrifugal action comes into play and causes an outward heel. This heel depends on several factors. Other things being equal it varies as the square of the speed, i.e. if the speed is doubled the heel is quadrupled. A vessel of small stiffness will heel more than a vessel of greater stiffness, and the smaller the diameter of the turning circle the greater is the angle of heel. In the *Lord Nelson* class for example the heel on turning during the first portion of the circle is eight degrees, this heel decreasing to five degrees as the speed is checked round the circular path. There is a distinct loss of speed on the circle owing to the fact that the ship's stern is pushed outside the circle by the force on the rudders, so that the thrust of the propellers is not in the line of motion. There is also the drag of the rudders. In one case the speed of $17\frac{1}{2}$ knots when putting the helm over was reduced to eight knots when on the circle, the engines still running at the revolutions for the $17\frac{1}{2}$ knots.

On the official steam trials of the ship, trials are made to test the capacity of each of the steering engines to take the rudders over when the vessel is

steaming at full speed and similar trials are made
with the vessel going astern at a lower speed. Trials
are also made with the hand gear and the time taken
to disconnect one steam steering engine and connect
up the other is tried as also the time taken to disconnect the steam gear and connect up the hand
gear.

Design of Rudders. It will be of interest to
consider briefly some points of importance in connection with the design of the rudders of a large,
fast war-vessel. We have already discussed the
reasons for adopting double rudders. Fig. 16,
showing a transverse section through the rudders,
has been drawn to illustrate these remarks. The
rudders must be completely under water and under
the protection of the protective deck, and no portion
of the rudders must be in contact with that deck,
because of the risk of injury if the deck were struck.
With an underhung rudder the bending moment on
the stock when put hard over at full speed is very
considerable, and the diameter of the stock must be
made of sufficient size to withstand this bending
moment. The bottom of the ship in way of the
rudders is formed of a steel casting, *A*, which embraces
both rudders and which is well connected to the
bottom plating of the ship which surrounds it. A
metal (phosphor bronze) casting, *C*, is fitted in the
casting at each rudder, and the rudder-stock is

enclosed by a metal sleeve.   Thus the bearing of the
rudder in the stern-post is metal to metal.   The hole

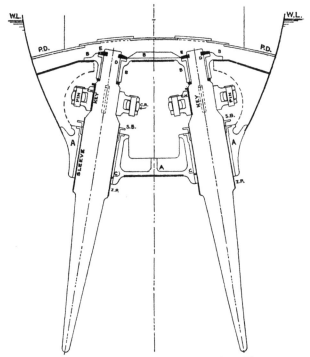

Fig. 16.   Transverse section through twin rudders.

in which the rudder works must be watertight, and
this is obtained by a stuffing-box, *SB*, at the upper

end. This stuffing-box is screwed down and compresses packing placed in the space between the metal casting C and the sleeve on the rudder-stock. The rudder is turned by means of a crosshead (CH) keyed to it and connected to the screw steering gear by means of connecting-rods (CR). The weight of the rudders is taken at the top. A steel casting, B, is fitted and supported by a massive structure going right across the ship. In the casting B are fitted metal castings, D. The head of the rudder is slotted out and a metal casting, E (fitted in two halves), is fitted and keyed to the rudder, and this bears on the metal casting D, proper provision being made for lubricating the bearing surfaces. A ring in two halves is fitted immediately under the casting B to prevent the rudder from lifting. Careful attention is necessary in the design of the rudder and stern-post to make sure that the rudder can be taken out when the vessel is in dry dock. In the case given in fig. 16 the rudders are inclined inwards at their upper end, this being necessary in order to obtain room for the crossheads. In ships not so fine at the stern as in this case, the rudders are fitted vertically. Zinc protecting plates (ZP) are secured to the rudder-stock just below the sleeve. When steel and a metal like phosphor bronze are in contact in sea water a galvanic action is set up, which corrodes the steel; but if zinc is fitted this corrodes instead of the steel, and can be periodically examined and renewed whenever the vessel is put into dry dock.

CHAPTER XI

POWER AND SPEED

Resistance. The resistance offered by the water to a vessel's progress is of two kinds, viz. frictional and wave-making. Frictional resistance due to the rubbing action of the water on the surface will depend primarily on the condition of the surface, and for this reason alone it is essential to keep ships' bottoms free from scale and weeds by frequent docking for cleaning and painting. The laws of frictional resistance are well known owing to the experiments of the late Mr W. Froude (after whom the new experimental tank at the National Physical Laboratory has been named). He experimented with a large number of boards of varying surfaces and at different speeds. The resistance is found to depend (1) on the nature of the surfaces, (2) on the area of the surfaces, (3) on the 1·83 power of the speed v and (4) on the length of the surface in the direction of motion. Expressed in a formula this is $R = f \cdot s \cdot v^{1.83}$. Other things being equal, to double the speed means increasing the resistance due to friction to $3\frac{1}{2}$ times its former amount. The value of the coefficient f increases very rapidly with the roughness of the surface. The coefficient f also decreases as the length increases, this being due to the fact that the rubbing induces a forward current of water and the rear of the surface is moving through water having a slight onward motion. This has to be taken into account when comparing the resistance of a model with a full sized ship as will be seen later. Resistance due to wave making is not so easy to understand. We have seen that frictional resistance does increase rapidly with the speed but wave-making resistance increases much more rapidly and this is the reason why high speeds in full sized ships are so difficult to obtain. When a vessel is towed through the water there are two separate and distinct series of waves created, viz. one series at the bow and one at the stern. These can sometimes be distinctly seen when a vessel proceeds in one of the Highland Lochs after a calm night with the surface of the water perfectly smooth. Each of these series consists of (1) a diverging series with the crests of the waves sloping aft and (ii) a transverse series with the crests nearly perpendicular to the centre line of the ship. The diverging waves at once pass away from the ship but the energy required for their formation is a portion of the wave-making resistance. The bow transverse wave passes along the side of the ship and at low speeds soon becomes

dissipated and indeed hardly noticeable. But as the
speed increases this wave becomes longer and higher,
the length increasing with the square of the speed.
The energy required to create continually such a
wave varies as the 4th power of the speed, i.e. if the
speed is doubled the energy of the wave system
increases to 16 times its former amount. When,
however, a crest of the transverse bow-wave system
coincides with the crest of the transverse stern-wave
system a very great increase of resistance results.
There is a definite relation between the length of a
wave from crest to crest and its speed, viz.

$$V = 1\cdot33\ \sqrt{L}\quad (V \text{ in knots, } L \text{ in feet}).$$

When the length of the wave is approaching the
length of the ship the wave-making resistance is increasing most rapidly because the crest of the bow-
wave series is coinciding with the stern-wave. This is
why the relation of the speed and the length is a
measure of the speed, whether high or low, for a
particular ship. When $V \div \sqrt{L}$ is 0·5 to 0·7 we have
a moderate economical speed; when $V \div \sqrt{L}$ is 0·7 to
1·0 we have the speed of mail-steamers and battleships; when $V \div \sqrt{L} = 1\cdot0$ to 1·3 we have the high
speeds of cruisers. Thus 15 knots is a high speed for
a vessel of 150 feet long but a moderate economical
speed for a vessel 500 feet long, the respective values
of the ratio $V \div \sqrt{L}$ being 1·22 and 0·67. In this

connection it is interesting to note that in the fastest
Atlantic Liners this ratio has been kept fairly constant
by increases in the length as speeds have been increased.

Etruria	19·5 knots	500 feet length	$V \div \sqrt{L}$	0·87
Teutonic	20 ,,	566 ,,	,,	0·84
Campania	22 ,,	600 ,,	,,	0·9
Deutschland	23¼ ,,	666 ,,	,,	0·9
Lusitania	25 ,,	760 ,,	,,	0·91

The same thing has been true in cruisers of high
speed.

Edgar	20·5 knots	360 feet length	$V \div \sqrt{L}$	1·08
Drake	23 ,,	500 ,,	,,	1·03
Monmouth	23 ,,	440 ,,	,,	1·1
Newcastle	25 ,,	430 ,,	,,	1·2
Invincible	25 ,,	530 ,,	,,	1·09
Lion	28 ,,	660 ,,	,,	1·09

In these ships the speeds are high and this necessi-
tates a great expenditure of power. In the battleships the speed is more moderate, e.g. $V \div \sqrt{L}$ is 0·95
in the *Dreadnought* of 21 knots and 490 feet length,
and 0·89 in the *Lord Nelson* of 18 knots and 410 feet
length.

If the speed is such that the accompanying wave
is longer than the length of the ship we reach a condition of things in which although the resistance is
high yet it is not increasing at so great a rate, and
further increments of speed can be obtained with

comparatively moderate additions of power. This is
only possible in vessels of the motor and destroyer
type in which very light fast-running machinery can
be fitted. The following figures for a destroyer were
given by Sir W. H. White in 1899 before the British
Association:

Up to 11 knots resistance varied as 2nd power nearly of the speed				
At 16	,,	,,	,,	3rd ,,
From 18–20	,,	,,	,,	3·3rd ,,
At 22	,,	,,	,,	2·7th ,,
At 25	,,	,,	,,	2nd ,,
At 30	,,	,,	,,	2nd ,,

The maximum rate of increase was from 18 to 20
knots and here the length of the accompanying wave
approximated to the length of the ship. At the higher
speeds the wave was longer than the length of ship
and the vessel was lying on the slope of a wave of
her own creation.

An important feature of the design of a vessel is
the form of the under-water body, and what is aimed
at is the form of least resistance. In addition to the
influence of length referred to above it is necessary
that the fore end shall be fine, especially at the waterline as if the fore end is bluff, large waves are created.
It is known from experiment and experience that to
minimise wave resistance in full-sized ships it is
necessary to dispose the displacement amidships by
having a full midship section with fine ends. It is

also of the greatest importance to make the run of
the ship fine so as to give a clear run of water to the
propellers and to avoid any eddying water at the
stern.   The first essential, however, to be satisfied in
the design of the form is the provision of a proper
amount of stability, and the question of trim also has
to be considered.   The designer will usually make
out a set of lines giving the form of the ship, which is
then sent to the Haslar Experimental tank.   An
improvement may or may not be possible.   A model
in paraffin wax is made and tried and the model
is run at speeds which *correspond* to the full-sized
ship.   *Corresponding speeds* are those which are
proportioned to the square root of the ratio of the
lengths.   Thus for a vessel of 480 feet long of which
the model is 12 feet long, the ratio is $\sqrt{40} = 6\cdot32$ and
23 knots for the ship *corresponds* to $23 \div 6\cdot32 = 3\cdot64$
knots for the model and similarly for other speeds.
A curve of resistance on a base of speed for the model
is then obtained by running the model in the tank.
The frictional resistance of the model is calculated
for the various speeds by the formula given above,
using an appropriate coefficient f for the length.
The difference between this and the total resistance
will give the wave-making resistance of the model for
the various speeds.   Now we can pass from the wave-
making resistance of the model to that of the ship by
a most important law enunciated by the late Mr W.

Froude called the *law of comparison* which is as follows:

If the linear dimensions of the ship be n times those of the model and the resistances of the latter at speeds V_1, V_2, etc. be R_1, R_2, etc. then the resistances of the ship at the corresponding speeds $V_1 \sqrt{n}$, $V_2 \sqrt{n}$ etc. will be $R_1 n^3$, $R_2 n^3$, etc.

This will enable the wave-making resistances of the full-sized ship to be calculated from those of the model at corresponding speeds. To these are added the calculated frictional resistances, using an appropriate coefficient f for the length and thus a curve of total resistance for the ship can be obtained on a base of speed.

The work done in towing a ship is the product of the resistance and the speed, and 33,000 foot lbs. of work performed in one minute is defined as one horse-power. This horse-power, which would be transmitted through the tow rope, is termed effective or tow-rope horse-power (E.H.P.), and is expressed in the formula.

$$\text{E. H. P.} = \frac{1}{326} \cdot R \cdot V \quad (R \text{ in lbs. } V \text{ in knots).}$$

In this way a curve of E.H.P. on base of speed can be obtained for the ship. This, however, is not the horse-power necessary for propulsion. The power transmitted through the shafts is wasted to a large extent by losses at the propellers, and from previous experience values of the ratio between the E.H.P. and the Shaft Horse-Power (S.H.P.) are known. This ratio is termed the propulsive coefficient and a usual value is 50 %. That is to say that one half of the power is wasted so far as the propulsion of the ship is concerned. The main causes of this loss are as follows :

1. The propeller blades experience edgewise and frictional resistance as they revolve through the water.

2. The presence of the propellers at the stern disturbs the even flow of water round the stern which otherwise would help the ship along. This causes an augment of resistance or may be regarded as a deduction of thrust.

3. The action of a propeller in revolving drives the water astern and it is the reaction which provides the forward thrust. The work of setting this water in motion is an unavoidable loss. There is also the circumferential motion given to the water which is a total loss.

Turbine machinery to work at its best should revolve at high revolutions but a propeller if rotated fast is inefficient. Thus a compromise has to be effected, viz. : the turbine is slowed down to help the efficiency of the propeller and the most efficient combination obtained. Very special care is taken in the design of propellers in regard to the thickness of the blades to reduce the edgewise resistance to a

minimum and special bronzes are employed which do
not corrode in sea water so as to reduce the frictional
resistance as much as possible.

It is a very difficult matter to settle offhand what
propellers will be most efficient for a given ship and
sometimes the trial results will indicate the direction
in which an alteration may be made with advantage.
For instance the *Drake* obtained a speed of 23 knots
with one set of propellers. The trials shewed that
there was excessive slip and indicated that the blade
area was insufficient. New propellers were fitted
which increased the speed to 24 knots for the same
power.

In this connection it may be interesting to refer to
Sir Chas. Parsons' method of geared turbines which
has been successfully tried in the cross channel steamer
Normannia belonging to the London and South
Western Railway and which is being tried in the
Royal Navy for some destroyers. In this system the
turbines are run at revolutions corresponding to their
maximum efficiency and a small pinion on the turbine
shaft is geared to a larger spur wheel on the propeller
shaft, the proportions of the gearing being such that
the propeller is rotated at a speed of revolution
corresponding to its maximum efficiency. It has been
found that by running the gearing in an oil bath and
paying special attention to the accuracy of the cutting
of the teeth and to the material of the helical spur

wheels the wear over long periods is negligible and the noise is quite slight.

Steam Trials. The trials carried out in ships of the Royal Navy, to satisfy contract conditions, are for *power* and not *speed*. (An exception is made in the case of some torpedo boat destroyers where the contract is for speed under specified conditions as regards loading and stability.) A certain specified power must be maintained for a given time and the speed obtained is the responsibility of the designer. To obtain an accurate measure of the speed the ship is run over a 'measured mile,' of which there are several round the coast, the best being at Skelmorlie near the mouth of the Clyde, where the water is deep and the course sheltered, and this course is used whenever possible. The measured mile consists of two pairs of posts erected on the land perpendicular to the course and exactly a knot (6080 feet) apart. The ship is brought on to the course and when the first pair of posts are in line a chronometer stop watch is started ; when the second pair of posts are in line the watch is stopped and the time taken is noted which can at once be turned into speed. In order to eliminate the effect of tide a number of runs are taken in each direction, six altogether being usual in modern ships. If the speed of the tide were uniform the ordinary average would give the correct mean speed, but this is not usually the case and what is termed the 'mean of means' is taken which will give the real mean speed. Thus if the speeds at successive runs up and down at equal intervals of time are V_1, V_2, V_3, V_4 then we proceed as follows:

Speeds	First means	Second means	Mean of means
V_1			
	$\frac{1}{2}(V_1 + V_2)$		
V_2		$\frac{1}{4}(V_1 + 2V_2 + V_3)$	
	$\frac{1}{2}(V_2 + V_3)$		$\frac{1}{8}(V_1 + 3V_2 + 3V_3 + V_4)$
V_3		$\frac{1}{4}(V_2 + 2V_3 + V_4)$	
	$\frac{1}{2}(V_3 + V_4)$		
V_4			

Similarly if six runs are taken up and down at equal intervals of time the mean of means is

$$\tfrac{1}{32}(V_1 + 5V_2 + 10V_3 + 10V_4 + 5V_5 + V_6).$$

(It would be out of place in a work of this character to explain the theory on which this method is based. This is given in works on Naval Architecture.)

While the vessel is running on each mile, records are taken of the shaft horse-power and the revolutions. For some ships, usually at least one of each class, measured mile trials are run for a series of speeds from about 10 or 12 knots up to the maximum. Such trials are called progressive speed trials and those of the *Gloucester* referred to later afford a specimen case. The information obtained when carefully analysed is found to be of the highest value especially in regard to the performance of the propellers, and the records of the speeds obtained by varying the revolutions are of great use in the subsequent navigation of the ship.

The steam trials for measuring the speed are carried out with the greatest possible accuracy. The ship is ballasted to keep her at the same draught for each series and the trials are run with the ship newly out of dock with the bottom freshly painted and in perfect condition. It is desirable that the weather should be favourable and the best coal, most skilful stoking and supervision are essential. Everything should be at the very best as otherwise we can have no exact knowledge of the performance as the influence of any factors affecting the speed of the ship adversely is unknown. It is not unusual however to find, in the subsequent history of a ship, that when the machinery has got into perfect working order, better performances have been made than even under the trial conditions.

The horse-power for reciprocating machinery is obtained from indicator diagrams taken from each cylinder from which the amount of work done in each for one stroke can be calculated and thence by bringing in the revolutions the power can be obtained. There is no corresponding method in the case of the turbine and the power is measured at each of the shafts. A shaft when transmitting power undergoes a slight twist and instruments have been devised for measuring this twist over a certain definite length.

Each shaft is tried in the shops to ascertain the amount of the twisting moment necessary to obtain a certain degree of twist. On trial the twist is measured and can be turned into twisting moment and thence by knowing the revolutions the power can be determined.

The results of the trials of the *Gloucester*, a second class cruiser 430 ft. long, 47 ft. beam and 4800 tons displacement, which have been published, illustrate very clearly the rapid increase of power with speed. The vessel was designed for a speed of 25 knots with 22,000 shaft horse-power. The trials on the measured mile gave the following results, viz. :

Speed in knots	Horse-Power
12·84	2,053
17·55	5,513
20·8	9,400
23·45	13,968
25·08	18,983
26·30	24,335

These results are shewn graphically in fig. 17, the values of the speed being measured along the base line and ordinates set up equal to the power, thus giving us a curve of horse-power as a speed base. Measurements from this curve give the powers at intervals of two knots from which the increment of power for every two knots is obtained. These are set out in the following table together with the value of the ratio $V \div \sqrt{L}$ at the various speeds.

Speed in knots	$V \div \sqrt{L}$	Horse-Power	Increment of Power for two knots
12	0·58	1,700	
14	0·67	2,600	900
16	0·77	4,100	1,500
18	0·865	6,000	1,900
20	0·96	8,300	2,300
22	1·06	11,300	3,000
24	1·15	15,500	4,200
26	1·25	23,000	7,500

It will be noticed how very steep the curve is at the higher speeds, indicating the great increase of power required for successive increments of speed. *To increase the speed from* 24 *to* 26 *knots requires as much additional power as would drive the vessel about* 19½ *knots, and to increase the speed from* 22 *to* 26 *knots means doubling the power.* The reason of this is of course the rapid growth of the wave-making resistance at speeds which are high for the ship in relation to her length as is shewn by the values of the ratio $V \div \sqrt{L}$ given in the table. The approximate curve of horse-power due to surface friction is shewn in dotted lines in fig 17. This gives some idea of the relative importance of friction and wave making at various speeds.

Effect of shallow water on speed. Attention has been directed in recent years to the important effect which depth of water has on the speed of ships. It was raised in a very acute form in the case of some destroyers which were found to vary considerably in

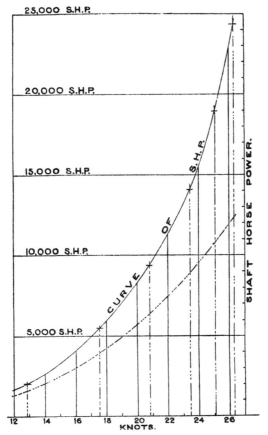

Fig. 17. Curves of Horse-Power as base of speed.

the speed obtained for the same horse-power with varying depths of water, on the measured mile. In one case 9000 horse-power gave 22 knots on the Maplins at a depth of water of $7\frac{1}{2}$ fathoms while the same power gave 26 knots at Skelmorlie at 40 fathoms, while 18,000 horse-power gave $34\frac{1}{2}$ knots on the Maplins and 33 knots at Skelmorlie. In the shallow water (at speeds below 26 knots in this case) the natural course of the stream lines round the ship becomes restricted owing to the nearness of the bottom and this causes a considerable increase of resistance. At the higher speeds however in shallow water (above 27 knots in this case) the conditions change and what is known as a solitary wave of translation is formed which moves forward with the ship, the after wave-system being wiped out with a consequent smaller resistance as compared with deep water. It is thus seen that to obtain proper results, steam trials, run for the purpose of speed records, should be run in water that is *deep* for the size and speed of the ship. If in moderately shallow water the speed will be defective and if in very shallow water it will be excessive as compared with the speed in deep water and consequently fictitious as regards the performance of the ship at sea.

A conspicuous instance of the effect of shallow water on speed was observed in the trials of H.M.S. *Edgar*. At Stokes Bay, where the depth of water is

12 fathoms, 12,260 H.P. gave a speed of 20½ knots. This result was disappointing, and on a trial on the deep sea course between Falmouth and Plymouth, at a depth of water of 30 fathoms, 12,550 H.P. gave 21 knots or about ¾ knot more for the same power. The Elswick firm once built a cruiser which obtained a higher speed on service than could be obtained on the measured mile before delivery. Subsequent trials confirmed the view that the reason for the deficiency in speed was the shallowness of water over the course originally used. It is on record that one shipbuilding firm was actually mulcted in a penalty because a certain contract speed was not obtained on trial and the ship on service from port to port averaged a better speed than that contracted for, the reason being subsequently found in the shallow water of the measured mile employed.

Similar results have been found to obtain in the United States Navy and the trials for large ships are now run on the Rockland course with a depth of 66 fathoms. The following figures have been published for the U.S. battleship *Michigan*, a vessel 450 ft. long, 80 ft. broad, 24½ ft. draught and 16,000 tons displacement.

	Measured mile Delaware 24 fathoms	Measured mile Rockland 66 fathoms
I.H.P. for 12 knots	3,525	3,300
,, 15 ,,	6,900	6,690
,, 18 ,,	13,525	13,325
,, 19 ,,	18,300	16,880

Or the power that gave 19 knots in 24 fathoms would give $19\frac{1}{4}$ to $19\frac{1}{2}$ knots in 66 fathoms of water.

The deepest measured mile course in this country is at Skelmorlie, near the mouth of the Clyde, where the depth of water is 40 fathoms or 240 feet, and this course is used whenever possible. Complaints, however, are being made as to the effect of the wash of water on the foreshore and it is very probable that in the near future important trials for speed records for large fast ships will have to be carried out on a deep sea course over a known distance. For instance the *Lusitania* not only ran a series of progressive speed trials at Skelmorlie but also had six runs between Ailsa Craig and Holy Island when the Skelmorlie results were closely confirmed.

Resistance of a vessel when fully submerged. A vessel wholly submerged like a submarine will not experience any wave-making resistance. The following table gives the E.H.P. of a submarine when on the surface and when submerged so that the centre line is about $1\frac{1}{2}$ times the diameter below the surface. The displacement and wetted surface when submerged are increased about 25 % and 15 % respectively as compared with the surface condition, but in spite of the added frictional resistance the E.H.P. is very much less.

Speed in knots	E.H.P. in surface condition	E.H.P. when submerged
8	60	35
9	100	60
10	190	80
11	210	100
12	260	170

The greater E.H.P. required for a given speed on the surface as compared with that submerged is due of course to the formation of waves which require energy to create them and are thus a cause of resistance.

CHAPTER XII

THE COST OF WARSHIPS

THE prices paid out for warship work are periodically published in the Navy Estimates presented to the House of Commons. The costs for hull and armour are grouped together so that information as to the separate costs is not available. The costs of gun mountings, torpedo tubes, propelling and auxiliary machinery and guns are given separately in detail. There is reason to believe that the firms contracting for armour and gun mountings make considerable profits on this work, but it has to be remembered that large capital expenditure is necessary and great risks have to be run especially in the armour plate industry. It has always been the policy of the British Admiralty to encourage private enterprise for this work rather than to undertake the work in Government workshops. We have already seen how new processes have been introduced in the manufacture of armour, each change involving great expenditure in experimental work and in new plant and machinery.

For hulls and propelling machinery the competition to obtain Admiralty work is exceedingly keen and there is reason to believe that work has been often taken in recent years at a very low margin of profit and possibly in some cases at an actual loss. 'Owing to great developments in the productive power of British warship building, competition not long ago reached a point when eminent firms are known to have made quotations which not merely included no profit but did not cover the whole of their establishment charges' (Sir W. H. White, *Naval Annual*, 1912).

The number of firms who are able to undertake large warship work is now very considerable, viz.:

Palmer's Shipbuilding and Iron Co., Ltd.   Jarrow-on-Tyne.
Cammell, Laird and Co., Ltd.   Birkenhead.
Vickers, Ltd.   Barrow-in-Furness.
Scott and Co., Ltd.   Greenock.
Fairfield Co., Ltd.   Govan.
John Brown and Co., Ltd.   Clydebank.
Beardmore and Co., Ltd.   Dalmuir.

Armstrong, Whitworth and Co., Ltd. Elswick (hulls only).
Swan Hunter and Wigham Richardson, Ltd. Wallsend (hulls only).
Hawthorn, Leslie and Co., Ltd. Hebburn-on-Tyne (machinery only).
Parsons Marine Turbine Co., Ltd. Wallsend (machinery only).
Wallsend Slipway Co., Ltd. Wallsend (machinery only).
Harland and Wolff, Ltd. Belfast (machinery only).

Thus, counting in the two Royal Dockyards at
Portsmouth and Devonport, there are no less than
eleven establishments where large war-vessels can be
constructed, and there are also eleven firms who can
undertake the machinery of these ships. Machinery
is not constructed in the Royal Dockyards.

Armstrong, Whitworth & Co., Ltd., and Vickers,
Ltd., are also constructors of large guns and gun
mountings and these firms together with the Coventry
Ordnance Co. have this work divided between them
with the exception of some gun work undertaken by
the Woolwich Government Arsenal and Beardmore
and Co., Ltd., at Parkhead. The firms of Armstrong,
Vickers, Brown, Cammell, John Brown and Beard-
mores are the makers of armour plates in this country.

The figures in the following tables have been
compiled from the published Navy Estimates. Table I
is a comparison of the three battleships *Neptune*,
Hercules and *Colossus* which for all practical purposes
may be regarded as sister ships.

It is seen that for these ships the armament of
10 12-inch guns, 16 4-inch guns and three torpedo

TABLE I.

	Neptune	Hercules	Colossus
Yard	Portsmouth	Palmer's Co.	Scott and Co.
Machinery	Harland and Wolff	,,	,,
Dimensions ...	510′ × 85′ × 27′	510′ × 85′ × 27′	
Displacement ...	19,900 tons	20,000 tons	
Shaft Horse-Power ...	25,000	25,000	
Speed. Knots ...	21	21	
Main Armament ...		10 12-in. 50 calibre	
Minor Armament ...		16 4-in.	
Torpedo Tubes ...	3–18 in.	3–21 in.	
Cost	£	£	£
Hull and Armour ...	829,760	845,166	845,953
Steam Boats ...	5,867	6,136	4,422
Machinery ...	250,063	237,281	252,525
Main Gun Mountings ...	382,103	394,569	391,778
Transferable ,, ...	12,673	13,231	13,263
Torpedo Tubes ...	3,860	5,294	5,405
Guns	131,700	131,700	131,700
Dockyard Establishment charges	89,932	27,573	27,614
Total cost exclusive of stores	£1,705,958	£1,660,950	£1,672,660

tubes costs about £545,000 or about 33 °/₀ of the total cost.

Table II gives particulars of the cost of the next class of battleship, the *Orion* class.

It is seen that for these ships the armament of 10 13·5-inch guns, 16 4-inch guns and three torpedo tubes costs about £588,000 or about 31 °/₀ of the total.

The gun mountings for the *Conqueror* are supplied by the Coventry Ordnance Co., a new competitor for this class of work.

The costs of the four vessels of the *King George V* class, the dimensions and particulars of which are given in 1912 *Naval Annual*, viz.: length 555 ft., breadth 89 ft., displacement 23,000 tons, shaft horse power 27,000, speed 21 knots, 10 13½-in. guns, 16 4-in. are as follows:

King George V. Portsmouth. Machinery by Parsons Co.	£1,961,096.
Centurion. Devonport. Machinery by Hawthorn, Leslie and Co.	£1,950,671.
Ajax. Hull and Machinery by Scott and Co., Greenock	£1,889,387.
Audacious. Hull and Machinery by Cammell, Laird and Co., Birkenhead	£1,918,813.

In these ships the cost of the armament is about £585,000 or 30 °/₀ of the total.

The battle-cruisers of the *Lion* class are the most costly vessels yet produced for the British Navy, the great speed of 28 knots (32 miles an hour) necessitating

TABLE II.

	Orion	Thunderer	Conqueror	Monarch
Yard	Portsmouth	Thames Iron Works	Beardmore & Co. Dalmuir	Armstrong
Machinery ...	Wallsend Slipway	,,	,,	Hawthorn, Leslie
Dimensions ...		545' × 88¼' × 27½' draught		
Displacement		22,500 tons		
Shaft Horse-Power		27,000		
Speed ...		21 knots		
Armament ...		10 13½-in. 16 4-in.		
Torpedo Tubes		3 21-in.		
Cost	£	£	£	£
Hull and Armour ...	973,053	1,008,667	1,037,254	1,034,001
Steam Boats ...	5,905	4,031	4,250	4,161
Machinery ...	266,043	255,763	251,781	234,747
Main Gun Mountings ...	421,892	423,338	374,379	416,780
Minor ,, ...	13,520	12,925	12,699	16,686
Torpedo Tubes ...	5,200	4,979	5,111	4,980
Guns	144,300	146,900	146,900	146,900
Dockyard Establishment charges	88,860	28,532	28,274	28,657
Total cost exclusive of stores	£1,918,773	£1,885,145	£1,860,648	£1,886,912

a large ship with 70,000 horse-power, the machinery
costing about half a million sterling. In Table III
the cost of the battle-cruiser *Queen Mary* has been
included; the particulars of this ship are as given in
the 1912 *Naval Annual* and are described as 'un-
certain.' These ships cost over two millions sterling.

The machinery of these vessels costs about 25 %
of the total and the armament about 23 % of the
total.

It is the policy of the Admiralty to require
delivery of these large ships in two years, or say 600
working days deducting Sundays and holidays. Out
of this time there has to be deducted the time for
docking and steam, gunnery and torpedo trials which
will possibly occupy thirty days, leaving about 570 days
to perform work costing in the aggregate two million
pounds sterling. The work is carried on in all parts
of the country but mainly in Sheffield, the North and
North-West of England, and Scotland, as even for a
dockyard built ship it is only the labour on the hull
that benefits the dockyard town.

As an illustration of the way in which Admiralty
work for hulls and machinery has been keenly com-
peted for by contracting firms, the figures in Table
IV have been compiled from the published Navy
Estimates for the protected second class cruisers
named after towns of Great Britain.

It is seen that the 5400-ton ships are costing

TABLE III.

	Lion	Princess Royal	Queen Mary
Yard	Devonport	Vickers	Palmers Co.
Machinery	Vickers	,,	J. Brown & Co.
Dimensions	660′ × 88¼′ × 28′ × 26,350 tons		660′ × 89′ × 28′ × 27,000 tons
Shaft Horse-Power ...	70,000		75,000
Designed Speed ...	28 knots		28 knots
Armament	8 13·5-in. 16 4-in.		8 13·5-in. 16 4-in.
Torpedo Tubes ...	2 21-in.		2 21-in.
Cost	£	£	£
Hull and Armour ...	992,237	1,003,774	1,058,800
Steam Boats ...	6,044	4,112	4,350
Machinery	498,603	516,566	505,803
Main Gun Mountings ...	317,665	317,831	320,133
Transferable ,, ...	12,504	16,710	16,315
Torpedo Tubes ...	4,446	3,831	4,632
Guns	118,300	120,300	say 120,300
Dockyard Establishment Charges	118,538	30,762	30,731
Total cost exclusive of stores	£2,068,337	£2,013,886	say £2,061,064

TABLE IV.

	Dimensions, etc.	Hull	Machinery	Total	Average
		£	£	£	£
Bristol	(1908–9 Estimates) —	168,142 (Brown)	154,491 (Brown)	322,633	
Glasgow	430′ × 47′ × 15′·3″	165,412 (Fairfield)	146,644 (Fairfield)	312,056	
Gloucester	4,800 tons	150,343 (Beardmore)	160,721 (Beardmore)	311,064	311,650
Liverpool	22,000 S.H.P.	151,116 (Vickers)	151,413 (Vickers)	302,529	
Newcastle	25 knots	156,332 (Armstrong)	153,636 (Wallsend Slipway)	309,968	
Dartmouth	(1909–10 Estimates) 430′ × 48½′ × 15′·6″	143,575 (Vickers)	137,766 (Vickers)	281,341	
Falmouth	5,250 tons	147,315 (Beardmore)	141,401 (Beardmore)	288,716	
Weymouth	22,000 S.H.P.	154,514 (Elswick)	184,375 (Parsons)	288,889	290,760
Yarmouth	24¾ knots	149,903 (London & Glasgow)	154,197 (London & Glasgow)	304,100	
Dublin	(1910–11 Estimates) 430′ × 49′·10″ × 15′·9″	153,050 (Beardmore)	135,313 (Beardmore)	288,363	
Southampton	5,400 tons 22,000 S.H.P. 24¾ knots	150,044 (Brown)	137,296 (Brown)	287,340	287,850
Chatham	—	Dockyard	135,409 (Thames Iron Works)	—	

slightly less than the 5250-ton ships and ships of
both these classes are costing considerably less than
the original ships of the type, of 4800 tons.   There
has been a steady diminution of the prices for the
machinery of constant horse-power.

BIBLIOGRAPHY

Article in *Encyclopædia Britannica*, by Sir Philip Watts, K.C.B.,
    F.R.S., on 'Shipbuilding.'
Text Book of Naval Architecture, by Prof. Welch, M.Sc.
Warships, by E. L. Attwood, R.C.N.C.
Shipyard Practice, by N. J. McDermaid, R.C.N.C.
Brassey's Naval Annual ⎫
Navy League Annual   ⎪ for particulars of war-vessels of all
Jane's Fighting Ships ⎬   countries.
Laird-Clowes' Pocket Book ⎭
The Marine Steam Engine, by R. Sennett and Admiral Sir H.
    Oram.

INDEX

Milton Keynes UK
Ingram Content Group UK Ltd.
UKHW041520181024
449640UK00009B/82